# Food For Thought

# Food For Thought

*An Epigenetic Guide to Wellness*

George J. Febish and Jo Anne Oxley

Copyright © 2011 by George J. Febish and Jo Anne Oxley.

Library of Congress Control Number:   2011911467
ISBN:         Hardcover            978-1-4628-4724-2
              Softcover            978-1-4628-4723-5
              Ebook                978-1-4628-4725-9

All rights reserved. No part of this book may be reproduced or transmitted in any form or by any means, electronic or mechanical, including photocopying, recording, or by any information storage and retrieval system, without permission in writing from the copyright owner.

This book was printed in the United States of America.

To order additional copies of this book, contact:
Xlibris Corporation
1-888-795-4274
www.Xlibris.com
Orders@Xlibris.com
102237

# Contents

Testimonials ..................................................................................17
Dedication ....................................................................................19
Acknowledgements.......................................................................21
Preface..........................................................................................23
    How the Book is Organized ....................................................24
    References in this Book ...........................................................25
Introduction.................................................................................27
    DNA-The Old Way of Thinking..............................................27
    The DNA Model .....................................................................28
    Epigenetics Definition..............................................................28
    Nutrigenomics Definition ........................................................30
        Cell Factories .....................................................................31
    Introduction Summary.............................................................31

Chapter 1: Early Life, Early Man and DNA .................................33
    Introduction.............................................................................33
    Basics .......................................................................................33
        Early Life on Earth............................................................34
        Natural Selection & DNA..................................................35
        Genome Project .................................................................36
        The Cell as a Computer.....................................................37
        More Americans are Growing Old than Ever Before..........38
        Early Man ..........................................................................38
    Advanced .................................................................................39
        Genes and Chromosomes...................................................39
    Science .....................................................................................41
    Conclusion ...............................................................................41
        Are We Victims of Our Genes ...........................................41

Chapter 2: Epigenetics ..................................................................43
    Introduction.............................................................................43

 Basics ................................................................................44
  We can Change Our Epigenetics ...............................44
  How does Epigenetics Work..........................................45
  Epigenetics is NOT Hard Wired ..................................48
  The Computer Analogy .................................................48
  Diseases can be Switched on by Epigenetics..............49
  Epigenetic Inheritance ...................................................51
  Epigenome .......................................................................52
 Science .................................................................................52
  Methyl Groups in Foods.................................................52
  Epigenetics in Science ....................................................53
  Cancer Research ..............................................................54
 Conclusion...........................................................................54
  The More we Know . . . The Less we Know .............54
  Understanding our Epigenetics.....................................55
  Largest ever Epigenetics Project Launched.................57

Chapter 3: Nutrigenomics..................................................58
 Introduction.........................................................................58
 Basics ....................................................................................58
  Disease and Nutrigenomics ...........................................58
  It starts in the Womb .....................................................59
 Advanced .............................................................................60
  The Nutrigenomics Organization (NuGO).................60
 Science .................................................................................61
  Prevention is Better than a Cure ..................................61
 Conclusion...........................................................................61

Chapter 4: Environment—You are what you Eat, Smoke and Drink.....63
 Introduction.........................................................................63
 Basics ....................................................................................64
  Our Food ..........................................................................64
  What has Changed in our Diets in last 11,000 years ....65
  You are what You eat .....................................................66
  Food labels .......................................................................69
  Food Additives ................................................................70
 Advanced .............................................................................71
  Food Quality....................................................................71
  America Got Fat..............................................................72

Protein .................................................................................73
　　Salt .....................................................................................74
　　Sugar .................................................................................74
　　Fruit Juices ........................................................................75
　　Grains ...............................................................................76
　　The Cooking Myth—"I don't have time!" ..........................76
　　Other Things We Ingest-The Smoking Syndrome ...............77
　　Food Related Chronic Illnesses ..........................................78
　Science ...................................................................................78
　　How Taste Works ...............................................................80
　Conclusion .............................................................................80
　　The Singing Scientist ..........................................................81

Chapter 5: Environment—You are what you Think ...................84
　Introduction ...........................................................................84
　Basics .....................................................................................85
　　Negative vs. Positive Thinking ............................................85
　Advanced ...............................................................................87
　　Forgiving the Unforgivable .................................................87
　　The Cost of Negative Thinking ...........................................89
　　Can Epigenetics build better Humans? ...............................90
　Science ...................................................................................90
　　What Science says about it .................................................90
　Conclusion .............................................................................91

Chapter 6: Environment—You are what you Believe .................93
　Introduction ...........................................................................93
　Basics .....................................................................................94
　　Religion .............................................................................94
　Advanced ...............................................................................95
　　Our Spirits and Life ............................................................95
　　Our Senses .........................................................................97
　　Man Made Sensory Control ...............................................98
　　God is all Loving ................................................................98
　　Physics, Good & Evil ..........................................................99
　　Religion—Food for Thought ..............................................99
　Science ...................................................................................99
　Conclusion ...........................................................................100

## Chapter 7: Vegetarianism ................................................102
### Introduction ................................................................102
### Basics ........................................................................103
#### Reasons for a Vegetarian Diet ..............................103
#### Milk, Apple Pie and Motherhood—The American Way ............104
#### Are Humans Carnivores? ...................................106
#### Human Dietary Change ....................................109
#### Healthy Eating ............................................111
#### Healthy Ethnic Foods .....................................112
#### The Nine Essential Amino Acids and Where to Find Them ......113
#### Vegetarians and Protein ..................................114
#### Table of Amount of Protein in many Foods ................115
#### Vegetarian Eating ........................................117
#### Health Reasons for Being a Vegetarian ..................119
### Advanced ..................................................................119
#### Statistics on Vegetarian Health .........................119
#### Cancer ....................................................120
#### Heart Disease ............................................120
#### High Blood Pressure .....................................120
#### Obesity/Body Weight .....................................120
#### Medical Costs ............................................121
#### Osteoporosis .............................................121
### Science ....................................................................121
#### History of Vegetarianism ................................122
### Conclusion .................................................................122

## Chapter 8: Glycemic Index ................................................126
### Introduction ..............................................................126
#### Table of Glycemic Index and load values138 .............127
### Basics .....................................................................129
#### Not all sugars are the same .............................129
### Advanced ..................................................................130
#### Diabetes ..................................................130
#### Blood Sugar Levels ......................................131
### Science ....................................................................131
### Conclusion .................................................................132

## Chapter 9: Take Control of your Thoughts, Actions and Diet ............ 133
### Introduction .................................................................................. 133
### Basics .......................................................................................... 133
#### Mass Misinformation ................................................................. 133
### Conclusion .................................................................................. 136

## Chapter 10: Don't let TV Advertisements Control your Actions .......... 138
### Introduction .................................................................................. 138
### Basics .......................................................................................... 138
#### Marketing ................................................................................... 138
#### Who gains from the ad? ............................................................. 139
#### Follow the Money Trail ............................................................. 140
### Conclusion .................................................................................. 141

## Chapter 11: America is Sick, Literally ................................................. 142
### Introduction .................................................................................. 142
### Basics .......................................................................................... 142
### Advanced .................................................................................... 143
#### Our Government's Impact on Cancer ......................................... 144
### Conclusion .................................................................................. 145
#### Do your own research ................................................................ 145
#### Develop a Plan for your Happiness,
  Health and Improve your Quality of Life ....................... 145

## Chapter 12: Drugs—Pro & Con ........................................................... 146
### Introduction .................................................................................. 146
### Basics .......................................................................................... 146
#### Our Environment ........................................................................ 146
#### Vitamins and Drug ..................................................................... 147
### Advanced .................................................................................... 149
#### Drug Companies ........................................................................ 149
#### Drug Research ............................................................................ 149
#### Medical Schools ......................................................................... 150
#### Drug Advertisements ................................................................. 150
#### Way of Life ................................................................................ 151
#### The Problem .............................................................................. 151
### Conclusion .................................................................................. 152

## Chapter 13: Sugar addiction ... 153
### Introduction ... 153
### Basics ... 153
#### Is Sugar an Addiction? ... 153
### Advanced ... 155
#### Adding Sugar to foods ... 155
#### We are all addicts ... 156
#### Breaking the Addiction ... 157
#### Breaking the Sugar Hold ... 159
### Science ... 160
### Conclusion ... 160

## Chapter 14: How do you get your Protein ... 161
### Introduction ... 161
### Basics ... 161
#### Prions can cause Disease ... 161
#### Does Protein in Food Trigger Epigenetics? ... 162
#### Ingested Protein ... 162
### Advanced ... 163
#### The Distance down the Evolutionary Chain ... 163
#### Fruits & Vegetables ... 163
#### Meats ... 163
#### Fish ... 164
#### Milk ... 164
### Science ... 164
### Conclusion ... 165

## Chapter 15: Diseases and Cures ... 166
### Introduction ... 166
### Basics ... 167
#### CDC Report on State of Aging ... 167
#### How to Prevent Aging Diseases ... 167
#### Quit ALL Tobacco Use ... 168
#### Regular Physical Activity ... 168
#### Good Nutrition ... 168
#### Increase Your Social and Mental Activity ... 168
#### Cost of Chronic Illness ... 169
#### Healthy Eating vs. Drugs ... 170

- Cancer.....170
- Cancer Treatments.....172
- Heart Disease.....172
- Diabetes.....173
  - What is Diabetes?.....173
  - What is Prediabetes?.....173
  - How does Diabetes affect Blood Glucose?.....174
  - How do the three main Types of Diabetes Differ?.....174
  - Complications of Diabetes:.....175
- Preventing Diseases.....175
  - Heart Disease.....175
  - Cancer.....176
  - Diabetes.....178
- Advanced & Science.....178
  - Body Signs.....178
  - Statistics.....179
- Conclusion.....179
  - Disease Solutions.....180

## Chapter 16: Epigenetics—Stress.....182
- Introduction.....182
- Basics.....182
  - Lowering Our Stress.....182
- Advanced.....185
  - Support Groups.....185
  - Changing Life Styles.....185
- Science.....186
- Conclusion.....186

## Chapter 17: Conclusion.....187
- Introduction.....187
- Basics.....187
  - It's your Health and Life at Risk.....187
  - Top Ideas to Remember from This Book.....189
  - Top Myths to Change in our Lives.....190
  - Future Directions.....190
- Losing Weight by Eating Healthy.....190
  - Dos.....191
  - DON'Ts.....192

  Sample Daily Menu .................................................................193
   All Day.....................................................................................193
   Breakfast .................................................................................193
   Morning Snack........................................................................193
   Lunch......................................................................................193
   Afternoon Snack......................................................................193
   Dinner.....................................................................................194
   Evening Snack.........................................................................194
   Meat Substitute Products .......................................................194
   Useful Cookbooks ..................................................................194
  Final Thoughts..........................................................................194

**Appendix A: Recommended Books, Videos & Web Sites** ....................197
  Books & Articles .......................................................................197
  Cookbooks ................................................................................198
  Videos on YouTube ...................................................................198
  Web Sites ...................................................................................199
  The Epigenetics Project BLOG ................................................200

**Appendix B: Thoughts to Remember** ................................................201

**Appendix C: Datamation Article in Paradigm Shift Column**.............203
  Does DNA Use Remote COM? ................................................203

**Appendix D: Powerful Ideas in this Book to Remember** .....................207
  Introduction..............................................................................207
  Chapter 1: Early Life, Early Man and DNA ..............................207
  Chapter 2: Epigenetics ..............................................................208
  Chapter 3: Nutrigenomics.........................................................208
  Chapter 4: Environment—You are what you Eat,
      Smoke and Drink................................................208
  Chapter 5: Environment—You are what you Think .................209
  Chapter 6: Environment—You are what you Believe ................209
  Chapter 7: Vegetarianism ..........................................................209
  Chapter 8: Glycemic Index........................................................209
  Chapter 9: Take Control of your Thoughts, Actions and Diet .......210
  Chapter 10: Don't let TV Advertisements Control your Actions....210
  Chapter 11: America is Sick, Literally........................................210

Chapter 12: Drugs—Pro & Con .................................................. 210
Chapter 13: Sugar addiction ...................................................... 211
Chapter 14: How do you get your Protein ................................. 211
Chapter 15: Diseases and Cures ................................................ 211
Chapter 16: Epigenetics—Stress ................................................ 211
Chapter 17: Conclusion ............................................................ 212
The BLOG .............................................................................. 212

**Appendix E: Top Ten Diseases** .................................................. 213
  Men's top 10 Disease Preventions ............................................ 213
    No.1—Heart disease ........................................................... 213
    No.2—Cancer ..................................................................... 214
    No.3—Injuries ................................................................... 214
    No.4—Stroke ..................................................................... 214
    No.5—COPD ...................................................................... 215
    No.6—Type 2 diabetes ....................................................... 215
    No.7—Flu ........................................................................... 215
    No.8—Suicide .................................................................... 215
    No.9—Kidney disease ........................................................ 216
    No. 10—Alzheimer's disease ............................................... 216
    Men's Summary .................................................................. 216
  Women's top 10 Disease Preventions ....................................... 217
    No.1—Heart disease ........................................................... 217
    No.2—Cancer ..................................................................... 217
    No.3—Stroke ..................................................................... 218
    No.4—COPD ...................................................................... 218
    No.5—Alzheimer's disease .................................................. 218
    No.6—Injuries ................................................................... 219
    No. 7—Type 2 diabetes ...................................................... 219
    No.8—Flu ........................................................................... 219
    No.9—Kidney disease ........................................................ 220
    No. 10—Blood Poisoning (Septicemia or Sepsis) ................. 220
    Women's Summary ............................................................. 220

**Appendix F: Famous Vegetarians** ............................................. 221
**Glossary of Terms** .................................................................. 223
**Foot Notes** ............................................................................. 225
**Index** ..................................................................................... 235

# Table of Figures

| | | |
|---|---|---|
| Figure 1: | Human cell statistics | 28 |
| Figure 2: | Chart of Early Man's Evolution | 35 |
| Figure 3: | Genes and Chromosomes | 40 |
| Figure 4: | DNA Replication | 46 |
| Figure 5: | Table of Common DNA | 47 |
| Figure 6: | Box of Rice Label | 70 |
| Figure 7: | Jar of Beans Label | 70 |
| Figure 8: | Table of Manmade vs. God Made Items | 81 |
| Figure 9: | Is the Glass Half Empty or Full? | 87 |
| Figure 10: | Religions | 95 |
| Figure 11: | The 5 Senses | 97 |
| Table 12: | Comparing Meat-Eaters, Herbivores and Humans (based on a chart by A.D. Andrews, Fit Food for Men, Chicago: American Hygiene Society, 1970) | 108 |
| Figure 13: | Table of Nine essential Amino Acids and Foods that Contain Them | 114 |
| Figure 14: | Table of Protein in Vegetables | 116 |
| Figure 15: | Table of Proteins in Beans | 117 |
| Figure 16: | Websites that list protein in different foods | 117 |
| Figure 17: | Our Typical Daily Protein intake | 123 |
| Figure 18: | Must Read Books & Cookbooks | 124 |
| Figure 19: | Glycemic Index Table | 129 |

**You Make Your Own Luck!
How to Change Your Health Luck!**

**We have more Control Than we Think!**

# Testimonials

**Food For Thought** made me seriously look at how my diet was affecting my health. I always knew diet was important but didn't realize it could change our genes and make us healthy or sick. My mother always said, "Eat your vegetables!"

—Kathy Fox, New Jersey, Project Manager

All my life I have eaten healthy. I enjoy a lot of different vegetables and eat some white meat and fish. **Food For Thought** has made me examine my diet even further.

—Clifford Sherman, Pennsylvania, Retired Accountant

Wow, we can control our own health! The thought alone is mind boggling. **Food For Thought** provides insights that go far beyond coming in out of the cold, or eating a balanced meal, or controlling calories. The science of epigenetics may be in its infancy but the tidbits of knowledge so far give great promise for its future.

—John Iannotti, Pennsylvania, retired manager/engineer

# Dedication

I dedicate this book to my fiancée, Jo Anne, who is my co-author, best friend and lover. Without her this book would not have happened.

# Acknowledgements

I like to acknowledge my business partner, David E.Y. Sarna who inspired me when we were writing the Datamation Articles. The last of these articles (in the appendix of this book) was my starting point on the journey that led to this book.

I want to thank the thousands of my BLOG readers (http://georgefebish.wordpress.com) that have inspired me to keep writing. Their interest and loyalty was a major motivator.

I also want to thank my family and Jo Anne's and also our friends and neighbors for all the questions asked. They made us think about what was important and how to present it.

# Preface

This is a book about us as humans, how we live, how our environment, thinking and eating can affect our well being and health. First let's discuss a little history on whom I am and where I came from as background. I have always been interested in why things work. This curiosity drove me to major in Physics in college. It was around the time that Biologists were talking about how DNA worked and mapping the Human Genome. I remember statements like "after we map the Human Genome, we will be able to know everything about life." Well it struck me back then that this was a lot of hope that probably would never come about. Life tends to throw curve balls at us all the time. The more we learn the more questions are raised and few answers are actually fulfilled.

I went through life always interested in DNA and how it was so much like a computer program. My field of work was with computers so as I learned more and more about them, I wondered how this model compared to our DNA. Later in my career, I had the opportunity to write a column with my business partner and friend, David E.Y. Sarna called The Paradigm Shift. It was in a computer magazine called Datamation. The column was designed to make people think about what was happening in computers at the time. We did this by being controversial. The magazine changed formats and became an internet periodical. Our last Paradigm Shift article was titled "Does DNA use Remote COM "(Component Object Model). The full article is in Appendix C. COM was a computer term describing a methodology for communicating between programs. It was very similar to what happened in the hardware industry when integrated circuits (chips) came about. You no longer had to build all the circuitry but instead would plug-in chips and interface them. This was much more efficient. The article went on to compare our DNA to computers communicating. It raised the question "If DNA is our hardware, where is the software located, and who wrote it?"

Although I didn't know it at the time, I was defining this biological software as epigenetics. Epigenetics is the basis of this book and is the relatively new area of biology that is re-writing all the text books. It defines a system that "plays" our DNA similar to how software "plays" hardware to achieve its purposes. I started to read everything I could find on epigenetics. Soon it became clear that epigenetics was a link between our environment and our DNA. As things changed in our environment (physical environment, food, thinking, etc), epigenetics "played" our DNA in different ways. Some of these are good and benefit us and others are bad and hurt us. Bad is a relative concept only from our point of view. From a life or DNA point of view, it's about experimenting with different conditions to move the genes forward and to SURVIVE another generation.

Later in my career I was diagnosed with high blood pressure and high cholesterol. I did not want to take drugs, so I read up on what caused these conditions and changed my lifestyle drastically. I became a vegetarian but ate some fish. I lost about 85 pounds and felt great. My next visit to the doctor revealed both my blood pressure and cholesterol were now low. This made me start to think about links between our health and mental well being and our DNA. All humans have basically the same DNA. So why do some of us get sick and others don't? Why does one twin of a set of identical twins get cancer and the other doesn't? Obviously it's not our DNA! Could epigenetics be playing their DNA differently?

I finally decided to write this book and describe what I have found in my research. I am not a biologist or a medical doctor. I am a guide that aggregates information from many different sources. This information allows you to make more informed decisions on your life. I highly recommend that you, after reading the book, investigate on your own. Then decide what changes, if any, you should make in your life. Live long, live happy, live healthy and improve the quality of the rest of your life. I love the saying, "the rest of your life begins today."

## *How the Book is Organized*

Most chapters in the book is organized as:
- **Introduction**—to the material in the chapter
- **Basics**—of the chapter for everyone to read

- **Advanced**—topics to be read by those that desire more information or a more detailed explanation of the subject
- **Science**—what science is saying about the topic and references
- **Summary**—conclusions drawn on each topic

At the end of the book is the final chapter, a conclusion of what was covered in the book. Following this conclusion are the various appendices and a glossary of terms.

## *References in this Book*

There are many facts in this book that come from the cited references. Read the referenced material and draw your own conclusions. There is far more material available to you than can be cited in any one book. The base of information is growing each day. Do your homework if you want to be healthy and have a good quality of life in your later years. It is in your hands only! Doing nothing, may very well make you sick.

In my BLOG, I am continuously updating the information presented in this book. Visit my BLOG at *http://georgefebish.wordpress.com*. Please enjoy the BLOGs there and leave comments. If you have stories of someone that saved their life or improved their health with diet, reducing stress or environment; please let me know on the BLOG.

Enjoy the book,
George J. Febish

# Introduction

The underlying concept of this book is based on two new fields: **Epigenetics** and **Nutrigenomics**. We will define these terms first and then use them throughout the book. The book is not a technical book but written for the average person on the street. Terms that are used a lot are defined in the text and other terms that you may not be familiar with are defined in the Glossary.

What are we made of? The lowest form of life is a cell. Every living thing is made of one or more cells. Multiple-celled creatures, like humans, are made up of colonies of cells. In complex creatures like ourselves, these colonies are divided into different types of cells and each has a specific function to do. The colony then grows in complexity and function. Humans have extremely complex functionality. How is this functionality and complexity controlled? Is our DNA everything? Does it control who we are? Does it control our health and mental attitude?

## *DNA-The Old Way of Thinking*

The common belief is all organisms have DNA. Over long periods of time this DNA mutates and changes. Species evolve into new species and old ones die out. The dinosaurs died out but they also evolved into our modern birds. Sex also causes two different DNA chains to merge and a new entity is created with modified DNA from the parents. Over time organisms evolved from one species to another. We evolved into current-day humans. Most people believe that DNA is like a hand in poker, you get dealt certain cards and you have to play with what you were dealt. Some of us were given "good" genes and others "bad" ones. In poker you can throw in some cards to get new ones but is this possible with our DNA? Can it be changed? Can it be controlled? Are we slaves to our DNA?

## The DNA Model

Our DNA is basically the same in every cell of our body. Cells have an outer work area and an inner nucleus that stores the DNA. Basically the DNA is used to build protein chains in the outer part of the cell. This protein is used to build things like the cell walls and to communicate with other cells or within the cell. DNA is subdivided into groups called genes. Genes are like small programs on a computer. When activated they build a particular protein chain that does something. Genes are further grouped into clusters called chromosomes (similar to libraries of programs on a computer). Genes are significant only when they are expressed (activated). An expressed gene does something while a deactivated gene is blocked from what it does. Activation is like a key fitting a keyhole and unlocking the lock. Deactivation is blocking the keyhole to prevent a key from unlocking it. Does it matter which genes are activated? Does it matter which ones are blocked or deactivated? What activates our genes?

| Number of cells in the human body | about 100 trillion cells |
| --- | --- |
| Number of chromosomes in humans | 46 (2 pairs of 23) |
| Number of genes in each human cell | estimated at 30,000—40,000 |

Figure 1: Human cell statistics

## Epigenetics Definition[1]

Epigenetics is a relatively new area of biology that is changing everything we thought we knew about life. The term epigenetics means "changes to the observable characteristic or trait of an organism. It causes gene expression by mechanisms other than changes in the underlying DNA sequence, hence the name epi—(Greek: over; above) genetics. These changes may remain through cell divisions for the remainder of the cell's life and may also last for multiple generations. However, there is no change in the underlying DNA sequence of the organism; instead, non-genetic factors cause the organism's genes to behave differently. This is similar to a stereo. It is a complex electronic device but it does nothing until someone turns it ON (expresses it).

The best example of epigenetic changes is the process of cellular differentiation. Stem cells become the various types of cells in our body. In other words, a single fertilized egg cell changes into the many cell types including neurons, muscle cells, epithelium, blood vessels etc. as it continues to divide. It does so by activating (expressing) some genes while inhibiting others."

This essentially means that there are things in each cell that activate or deactivate our DNA genes. These things come from a wide variety of influences including food, environment, our parents, grandparents, etc. We have many different cell types but they all have essentially the exact same DNA. What is different is the epigenetics. Some of our DNA is critical to human life and normal development. Some is old DNA from long ago ancestors. Epigenetics can be made to turn on important genes or turn off bad genes or old genes that shouldn't be turned on.

The cell doesn't know what is good or bad. It reacts to changes in its environment. These changes can cause the cell to activate or inactivate different genes in an attempt to adapt to the changes. The answer if it is a good change or a bad one can only be evaluated long afterwards. If it helped the cell and ultimately the organism to adapt and prosper to the changes, then it was a good reaction. If the cell or organism died then it was a bad reaction. Organisms that die don't get to pass on their DNA or epigenetics so that is a good thing.

Epigenetics and genetics are a memory system. *Bad ideas die and are forgotten. Good ideas are remembered and passed on to future generations.* Our DNA is very old from the beginning of life on Earth. Our DNA doesn't know the difference between good vs. bad but it remembers changes. Good changes cause this DNA combination to have a higher probability of survival. It then gets to be around in all life after that point. Bad changes have a higher probability of the DNA's life form dying out. It then doesn't get to be passed on or remembered. So our DNA is the sum result of all the correct decisions our DNA went through from its origin to humans. The problem is that not all the genes are currently needed or good for us. Some genes are good at early stages of fetus development but bad later in our life. Turning on these genes by accident may be catastrophic to our well being. Since what we interact with DNA is the changes that cause genes to be turned ON/OFF; we might be turning on the wrong genes and off the right ones. This is very complex and is a form of programming our genetics.

This ability to modify what genes are expressed (turned on) or not expressed (turned off) allowed us to evolve through changing climates and times. It caused us to develop from primal forms of life to mammals and on to humans. In the past for billions of years these changes were mostly brought about by environmental changes that caused us to evolve. In modern times, humans are causing most of our epigenetic changes. We are changing our environment, our food supply, and the stress that is on us. These changes are happening in a very short time frame and are causing drastic changes to the phenotype (any *observable characteristic* or trait of an organism). Not all these changes are for the good. We can see the result of this everyday by how many humans, especially in America, are sick, are seeing a psychiatrist, are having the quality of their life diminished.

We used to think that it was our genes that made us what we are but genes are pretty much hard wired and very similar in all humans. Now we know that it is a combination of our genes and our epigenetics that make us who we are. The epigenetic side is inherited from our parents and maybe even our grandparents but it is changeable. Don't say or think "I got the bad genes". You actually may have gotten the bad epigenetics but you can change it.

Two great examples of DNA vs. epigenetics are:

1. Our DNA is like a computer's hardware and epigenetics like its software. The environment (food, thinking and physical environment) is like data to the software. In a computer system the software uses the hardware instructions to accomplish its goals. It reads data and acts upon it to complete the original programmer's ideas. In our bodies epigenetics uses the DNA instructions (genes) to accomplish its goals. The environmental conditions are data that causes the epigenetics to meet nature's goals.
2. Our DNA is like a musical instrument and epigenetics is like the musician. The sheet music is like our environment. A musician reads the music and uses his or her skills to play the instrument (hardware). In our bodies epigenetics uses environmental conditions to play our DNA. Just like some musicians may play well and others may play badly; how we interact with our environment causes our epigenetics to either play well or poorly.

How can we tell what changes are good or bad for us?

## *Nutrigenomics Definition*[2]

Nutrigenomics is the study of the epigenetic effects of foods on genes. It looks at each expressed gene and which proteins they create to better understand the effects on our health. Nutrigenomics aims to develop rational means to optimize nutrition, with respect to the subject's genetic constitution of a cell, an organism, or an individual.

This area will become significant in the near future as science tries to better understand what causes our moods and health. There will be a massive study of epigenetics called the epigenome. This will try to link which proteins activate or deactivate which genes. Next Nutrigenomics will link the external food that causes some of these epigenetic changes.

**Cell Factories**

Each cell is a mini factory that needs raw materials as inputs. It has machinery like the ribosomes to build things and it produces an output which is protein. Some of this protein is used to build various things in the body and others are used to communicate with the other cells. Each small factory must stay in touch with the other small factories to ensure the job is completed correctly. Our DNA is then a manifesto of what and how the factories are to operate. It has sections that control different factories (the different cell types). Each factory (cell) looks at that part of the manifesto (DNA) that was written for it. If a factory suddenly stops getting a required raw material or the materials is changed substantially, it may not operate correctly. This causes it to produce inferior products. This condition in our bodies is called sickness.

## *Introduction Summary*

This book is not designed to sell you anything. It is designed to show you things that you need to do better research on. **ONLY YOU CAN MAKE THE QUALITY OF YOUR LIFE BETTER!** No one else is responsible for it except you. Wake up, read on and make some real changes in your life.

During the renaissance period, people had great powers of observation. They looked at how things worked and changed their lives to match it. We seem to have lost all our powers of observation. How can we be sold more and more drugs, bad food and see that America is getting sicker and sicker and doing nothing about it? If you care about your quality of life and that of your families, read on and change your life forever.

You can agree, partially agree or totally disagree with things in this book. There is one thing you must agree with: *Your health and happiness is your responsibility not your children's your parents', your neighbors', your doctors' or your psychiatrists'.* If that is the only thing you take away from this book, we have succeeded.

In this book, we have presented many facts and references to other books, articles and videos that will help you become and remain healthy. This is a comprehensive set of information not found, in a single volume, anywhere else.

Be like Renaissance Man was—study, investigate and make changes accordingly. You must firmly believe that only you can make your quality of your life better. It is really not in anyone else's interest to do so. After years of a poor quality of life, what do most people ask? "Why me, God?" It is also not God's fault.

# Chapter 1:
# Early Life, Early Man and DNA

## *Introduction*

As human beings we have evolved through some very hard times and survived. We have been around in one form or another for about 2 million years (the genus Homo is 1.5 to 2 million years old but Homo sapiens are about 150,000 to 200,000 years old). During that period we have evolved based on countless changes in our lives. Most of these changes happened slowly. Our bodies have evolved to handle problems of finding enough food, climate changes, stress of being attacked by predators and millions of other life effecting changes. Our early ancestors were probably eaten more than they ate other animals. Life was not always good. It was much different than it is today. We were not always on top of the food chain. Early man was opportunistic and found food where he could. He found fruits, berries, nuts and vegetables and dead animals. Sometimes he even caught animals. As he evolved and started to make weapons he got better at defending his group from other groups and wild animals. We are who we are today because of these changes.

## *Basics*

Our genetics (DNA) evolved slowly over long periods of time and was the result of a vast number of different species on Earth. Our epigenetics (details in the next chapter) changed very rapidly based on our food, our stress levels and our environment. Genetics are hardwired well proven conditions while epigenetics are shorter term reactions to our lives and what is occurring in real time. Together they make each of us who we are and what we will become. How did life get started and evolve into humans?

**Early Life on Earth**

Scientists do not agree on how life started on Earth but more of them believe it may have come to Earth like our water on meteors. First giant ice meteors crashed into the Earth delivering water to it. The oceans were formed and this gave an environment friendly to our kind of life. Then other meteors may have brought amino acids the building blocks of life. These small molecules are essential for life as we know it but are not living organisms. The amino acids chain link together to form various proteins. Simple forms of life like bacteria and viruses were first to form. Single cell organisms came next. Soon cells learned how to group together. Specialized tasks were done by different cells. This was the basis of all complex life on Earth. Mammals, from where we come, are one of the oldest complex life forms on Earth.

Cells became very intelligent in how they divided up their tasks, reproduced new offspring and communicated with each other. This was all under the control of genetics (DNA) and epigenetics. We will talk much more on these subjects in the book. DNA was moved from its simple non life components, amino acids, to complex cells working as an organism. Eventually man evolved and dominated the Earth.

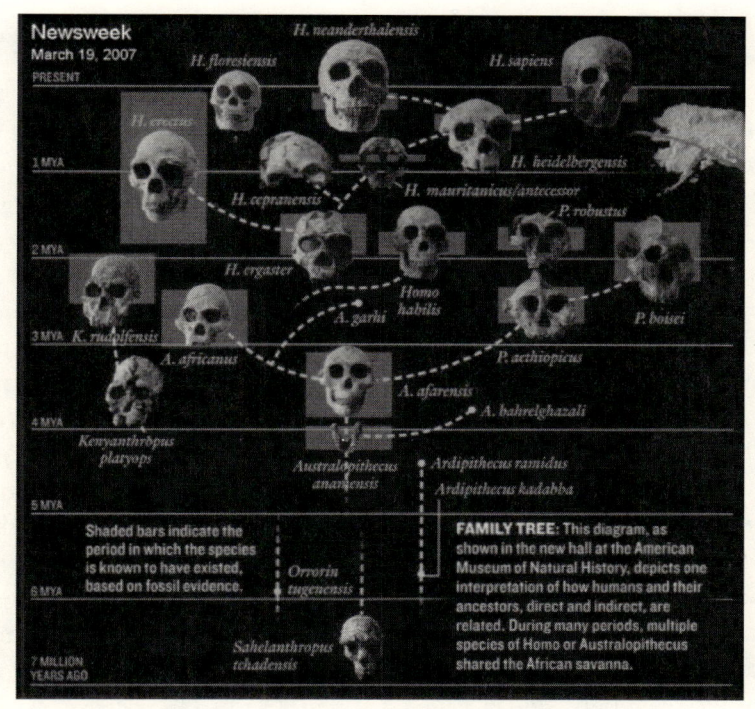

**Figure 2: Chart of Early Man's Evolution**[3]

## Natural Selection & DNA

Scientists used to believe humans evolved only by natural selection over very long periods of time. Basically our DNA changed for better or worse. Being sexual, man has been able to mix up two DNA's from a woman's egg and a man's sperm to form new life. The new fertilized egg has a new and different mix of DNA. Those that had *good* changes had an advantage and therefore a higher probability of procreating a new generation. It may sound harsh but DNA is designed for the advancement of itself not the species. If we wipe ourselves out some other species will dominate, become intelligent and move the DNA code forward. The dinosaurs were on top of the food chain and were becoming more intelligent until they were wiped out by a meteor and that cataclysmic event ended their DNA reign and opened up ours, the mammals. DNA is all about survival, opportunity and adaptation. The species capable of doing all three of these well, survives. How does DNA adapt and make changes to benefit us?

## Genome Project

Scientists wanted to determine how the cell worked. They knew DNA was the basis of heredity but how did it work? They soon learned that DNA is a memory system that basically creates various types of proteins needed by the body. Our DNA is grouped together in working modules called genes. These genes are like computer programs. When they are run, they do something. They thought if they could only know all the DNA gene codes they would understand life. This gave way to the Genome Project to map each human gene and determine what protein it builds. Very recently scientists completed this project and identified all the genes represented by our DNA. They thought they would know everything about life but they were in for another shock.

DNA is like the Intel Instruction Set on a PC. If we were to look at the instruction set we would see math functions like add and multiply; logic functions like AND OR NOR; Memory Functions like move, etc. That doesn't explain Windows, picture editing programs, video programs, word processors, spreadsheets etc. They are built from software that uses a combination of the hardware instruction set to implement the various programs we see when we use a PC. It is very hard to look from a detailed instruction set up to functionality. Looking at the Intel set in no way explains the programs that have been implemented on top of it. Going downward from this functionality to the underlying instruction set is easy. In software this is called *decoding*.

We believe that DNA is similar in concept. It is very hard to look at DNA and understand the various body functions and how they were built. There must then be something else that resides on top of our DNA similar to software that resides on top of the hardware. It is only recently that this was discovered.

Dr. Bruce Lipton, an American developmental biologist at New Zealand College, compares the cell membrane to a computer chip. This analogy of the programmable bio-computer sums up most of what this book is about:[4]

> The fact that the cell membrane and a computer chip are homologues means that it is both appropriate and instructive to better fathom the works of the cell by comparing it to a personal computer. The first big-deal insight that comes from such an exercise is that computers and cells are programmable. The second corollary insight is that the programmer lies outside

the computer/cell. Biological behavior and gene activity are dynamically linked to information from the environment, which is downloaded into the cell.

As I conjured up a bio-computer, I realized that the nucleus is simply a memory disk, a hard drive containing the DNA programs that encode the production of proteins. Let's call it the Double Helix Memory Disk. In your home computer you can insert such a memory disk containing a large number of specialized programs like word processing, graphics and spreadsheets. After you download those programs into active memory, you can remove the disk from the computer without interfering with the program that is running. When you remove the Double Helix Memory Disk by removing the nucleus, the work of the cellular protein machine goes on because the information that created the protein machine has already been downloaded.

Data is entered into the cell/computer via the membrane's receptors, which represent the cell's "keyboard." Receptors trigger the membrane's effectors proteins, which act as the cell/computer's "Central Processing Unit" (CPU). The "CPU" effectors proteins convert environmental information into the behavioral language of biology.

The real question is "What is this data that crosses the cell's membrane and affects its operation?" This is one of the main topics of this book.

## The Cell as a Computer

Our cells are individual living organisms that make decisions and extend their life through the creation of new cells. Dr. Lipton's idea of the membrane of the cell acts like computer code shows how programmable we really are. What are the data items that affect these tiny biological computers? It turns out that all environmental changes (food, chemicals, stress, etc.) affect us. They in turn trigger genes in the DNA code. The external world is data to the cell. It causes the cell to do different things in reaction to differing data inputs. Some of these reactions are good and some are bad. The cell is not trying to aid the human but survive itself and produce a new generation of cells.

It is always a scary thought to think of humans as a computer that is programmed to react to its environment. The analogy is a good one which describes how and why our individual cells do what they do. We cannot always control everything in our environment but we can control a lot of it. Our cells have evolved over a long time and can handle a lot of changes even bad ones. Modern man has thrown some very drastic changes at his cells in the form of food supply, thinking, and environmental conditions. All of these changes are too much for the cells to handle properly causing us to get sick in response to the changes.

## More Americans are Growing Old than Ever Before

The really scary thing is we are living longer. "We are in the midst of a demographic shift unprecedented in history", said Dr. Richard J. Hodges, former director of the National Institutes of Health in his 2005 budget request. [5] Dr. Hodges says at a time when there were 300 million Americans, there were 35 million older than 65, 4 million older than 85 and 65,000 celebrated their 100th birthday! If the rate of disease keeps at what it is today, we will have a national disaster on our hands. Higher rates of disease mean more hospitals and higher healthcare costs. We must lower the rate of illnesses and get our elderly back to a normal healthy life style. If we do not make changes, we will all pay for it in taxes.

### *Early Man*

Early man basically was a hunter gatherer. They searched for food and followed animals to water and occasionally tried to kill them for food. We imagine early man was hunted for food by other predators more than he hunted other animals. They would walk a lot, sprint at times and run as fast as they could when chased. Their evolving brains gave them the real advantage over just DNA and epigenetics. They could observe things, think about them and adapt how they would react. This in turn would modify their epigenes and pass them on to future generations.

Our bodies evolved a system that worked well then but not so well now. It would burn calories when it thought it was getting enough food and store fat and resist burning it when they were not getting food. So if food was plentiful, early man ate and ate probably small meals several times a day. Their body would register that food was coming in on a regular basis and turn off our "Starvation Mode" protection mechanism allowing unused food to be easily excreted as waste. But as early man started to move and may have hiked for days without finding much food, the "Starvation Mode" mechanism would activate via epigenetics and allow our stored fats to last longer. When they did eat any excess would be stored as fat. The mechanism saved early man from starvation on their quest.

Today, this mechanism is counter intuitive and often gets in our way. Food is usually available even though our bodies may think it is not. If you want to lose weight fast, the most obvious answer would be not to eat! This would be the wrong choice. It would enable that ancient "Starvation Mode" mechanism and our bodies would fight us to protect each pound of fat. If we were to eat small meals of fruit, vegetables and nuts several times a day, we would not be hungry and the "Starvation Mode" mechanism would switch off causing excess calories to be excreted as waste and any stored fat to be burned off. So to lose weight, you must eat! You must eat smart not stupidly and that is the trick!

## *Advanced*

**Genes and Chromosomes**

Our DNA is sub-divided into groups of instructions or programs that do things called genes. Our genes are packaged into 46 chromosomes (actually 23 unique ones since each chromosome is duplicated in the body—one set from our mother and the other set from our father). What are important from this books point of view are the genes. They do things to help build new cells, repair damage, etc. In fact they built each of us from 1 fertilized egg. Genes are like programs on a computer, not all of them are executed or run. Some are dormant others run only occasionally and still others run all the time. Genes can also be *locked* off. Like putting Parental Controls on certain TV channels so your kids can not watch them (it blocks them from viewing those channels).

Keeping with our computer analogy; some functions of genes are old and may be dormant or not needed anymore. These genes are not normally turned on but the data we spoke about above can affect them and turn them on. Your computer may have a lot of programs. Some may be really old and may no longer work. They cause no damage as long as they are not activated. Activate them and anything can happen. We now know that some genes can cause cancer but only if they are activated!

Cells receive some data that tells them their specialized function in the organism. This data turns on and off a lot of DNA functionality. They may be a heart cell, lung cell, blood cell, muscle cell, etc. Once assigned, they never change and all their offspring are the same functionally. A cuticle cell grows finger nails and heart cell does not. You would not want your heart growing nails would you?

The complex relationship between cells and the organism has been evolving for hundreds of millions of years maybe even billions of years. Each cell has the capability of doing it all. After all; all of our cells contain the same DNA. Just like human workers must specialize to get a job done; our cells must specialize for the good of the organism. If the organism survives, the cells survive. If the cells survive the DNA survives and has the opportunity to mutate or mix sexually to form new DNA combinations. Some of these combinations will result in problems and the organism will die out but others will give the organism an advantage and it will survive longer. Still other mutations will cause the organism to transgress into a new organism over long periods of time.

The important concept here is the data that influences a cell's operation. What is this data and where does it come from?

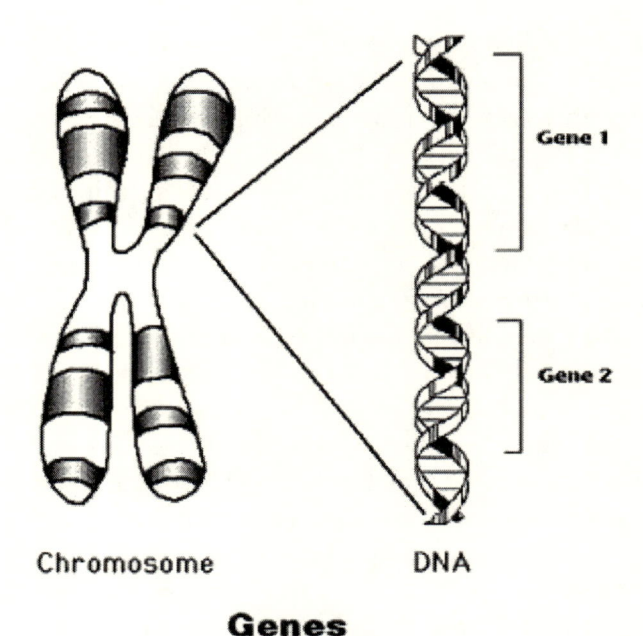

**Figure 3: Genes and Chromosomes**[6]

## *Science*

In 1997, the first successful extraction of Neanderthal DNA was announced.[7] A German & American team extracted mitochondrial DNA (mtDNA). So what is mtDNA? Our normal DNA also called "Nuclear DNA" is found in the nucleus of every cell. Mitochondria are small energy producing organelles found in cells. Mitochondria have their own DNA. Most cells contain between 500 & 1000 copies of the mtDNA. This makes finding it and extracting it a lot easier than nuclear DNA. Usually mtDNA come from our mother's egg not a combination of mother & father.

If we look at everyone alive today, we have a set of mothers that produced these offspring. This in turn had a smaller set of mothers and so on back until there is one common mother. All our mtDNA came from this common ancestor. This means that similarity of mtDNA for any two humans provides a rough estimate of how closely they are related through their maternal ancestors.

This is one tool scientists are using to track our ancestors and determine how much we have in common. The studies of Neanderthal mtDNA do not show that Neanderthals did not or could not interbreed with modern humans. However, there is a lack of diversity in Neanderthal mtDNA sequences, combined with the large differences between Neanderthal and modern human mtDNA. This strongly suggests that Neanderthals and modern humans developed separately, and did not form part of a single large interbreeding population. The Neanderthal mtDNA studies will strengthen the arguments of those scientists who claim that Neanderthals should be considered a separate species which did not significantly contribute to our modern gene pool.

## *Conclusion*

### Are We Victims of Our Genes

If we look at a gene as a computer programmed to respond in certain ways to change, the answer is no we are not victims to our genes. Computer programs written by man try to isolate bad inputs and deal with them but every once in a while bad data can cause the program to fail (death) or start to do things wrong (illness). We have said that the purpose of a gene is to build a protein (made of a sequence of amino acids) that is needed by the body. In reality a single gene can code for up to 30,000 different variations of a protein. Not all variations will have an effect on the cell. Some will have a positive effect and others a negative one. Dr. Bruce Lipton says:[8]

- Some variations may cause mutant characteristics that can lead to chronic illness and death.
- You are NOT trapped! You can rewrite your genetic expressions through your mind.
- Your genes pose NO limitations on your health or well being.
- Genes simply respond to data coming from our environment

Imagine our genes pose NO limitations on our health or well being! If this is true, what does cause us to be sick, to die prematurely or to be sad and depressed? It must be something outside or above the genes. If we find this influencing factor on our genes, can we control it?

It would appear that evolution has three parts:

1. The long term mutation type that mutates one life form into another (changes to base DNA),
2. Intermediate sexual propagation which mixes a mother and father's genes together to form a new life form, and
3. The immediate short term type that helps us to deal with the day to day changes in our lives. These triggers can be everything that interacts with us including food, environment, stress, thinking, etc.

The first two get hard wired in us as living creatures. The immediate type helps us use the hard wired DNA to react to immediate changes in our lives. Some of the immediate types are good for us and some are bad. Our individual bodies have no way of knowing what is good or bad in advance. Each of us reacts in different ways. Some will suffer from the interaction and over time that gene pool (including its epigenetics) will die out. Others will overcome the changes and survive. Remember it is about survival of the genetics including both the DNA and the epigenetics that matters not any one species. Species will come and go as they have in history.

# Chapter 2:
# Epigenetics

## *Introduction*

Epigenetics is a relatively new area of modern biology. Every major university now teaches epigenetics and most have an epigenetic research department. This research is causing all the biology textbooks to be re-written. A Google search produced 2,690,000 hits in June of 2011. Although understanding our genes and DNA is important, understanding epigenetics is critical. A major function of this book is to bring an understanding of epigenetics to the general public. In a few years it will be a household term just like DNA is today.

In the last chapter we saw that our DNA and genes do NOT pose limitations on our health or well being. What does then? Our cells reacting to the environment they are in cause changes in which genes will be run or not run. This is epigenetics. The really good news is that epigenetics is NOT hard wired. It is changeable through our lifestyle. How we live our lives, become the music our epigenetics plays on our genes. What kind of music are you playing?

Epigenetics react to things that come from a wide variety of influences including food, environment, our parents, our mood, etc. Epigenetic's can be made to turn on important genes or turn off bad genes making us healthy and happy or to turn off important genes and turn on bad ones making us sick and depressed.

Epigenetics and genetics are a memory system. Our DNA is very old from the beginning of life on Earth. It doesn't know the difference between good and bad but it remembers changes. Good changes cause this DNA combination to have a higher probability of survival. It then gets to be around in all life after that point. Bad changes have a higher probability of the DNA's life form dying out. It then doesn't get to be passed on or remembered. So our DNA is the sum result of all the correct decisions our DNA went through from its origin to humans. The problem is that not all the genes are currently needed or good for us. Some genes are good at early stages of fetus development but bad later in our life. Turning on these genes by accident can be catastrophic to our well being. Since what we interact with is changing and causes genes to be turned ON/OFF; we might be turning on the wrong genes and off the right ones.

We now know that our genes don't predetermine who we are. Our epigenetics do! The epigenetic side is inherited from our parents and maybe even our grandparents but it is changeable. Don't say or think "I got the bad genes". You actually got the bad epigenetics but you can change it. Our habits, eating styles, how we think and what we believe all affect epigenetics. What we really need is a course on how to make positive changes to our lifestyle that will ultimately have a positive effect on our epigenetics and therefore our health and well being.

## *Basics*

We defined epigenetics in the Introduction to the book. Here we will take a deeper look at what it is and what it can do.

### We can Change Our Epigenetics

"Epigenetics introduces the concept of free will into our idea of genetics"[9], we are not dealt a fixed deck of cards. We can change the cards in our hand and get a better set for our quality of life and happiness. "People used to think that once your epigenetic code was laid down in early development, that was it for life," says McGill University epigenetic pioneer Moshe Szyf. "But life is changing all the time, and the epigenetic code that controls your DNA is turning out to be the mechanism through which we change along with it. Epigenetics tells us that little things in life can have an effect of great magnitude." "Inheritance" comes in many different forms: we inherit stable genes, but also alterable epigenes; we inherit languages, ideas, attitudes, but can also change them. We inherit an ecosystem, but can also change it. [10]

Can something as simple as environment affect our inner workings? The answer is YES! In an experiment with mice, scientists have found the proof. *Invisible differences effected mice in an experiment done in 1999, by Oregon neuroscientist John C. Crabbe. He did a study on mice on the effects of alcohol and cocaine. He found that mice with identical DNA and identical test environments located in different cities yielded different results.*[11] It seems that everything in this world interacts with us and each of us with it. Shenk says, *"Genes are constantly activated and deactivated by environmental stimuli, nutrition, hormones, nerve impulses and other genes[12]"* In fact everything we know about this world is from our five senses. Without any of the five, we would have no knowledge of life here at all. These five senses along with eating, temperature, sun light, etc. causes chemical changes in our bodies that affect the epigenes.

Victor McKusick, the John Hopkins geneticist widely regarded as the father of clinical medical genetics, reminds us that in some instances "two blue-eyed parents can produce children with brown eyes." Recessive genes cannot explain such an event; gene—environment interactions can. [13]

Think of our DNA as a musical instrument. All humans have pretty much the same physical instrument (DNA). We are what our epigenes *play* out on our gene base. Each of us has a unique song being played (even identical twins). That song determines how we live our lives, our intelligence and our quality of life. Anyone can pick up a flute and make noise with it. A musician with experience playing a flute can make beautiful music with it. All humans can play their DNA with their epigenetics. Some just make noise and will end up sick or depressed. Others learn to play it correctly and will remain happy and healthy throughout life. The good news is we can choose to play a different tune and change our lives. We must first want to play a different tune. This is the concept of free will. We really do have the power to play whatever tune we want. Which do you want? Are you just making noise or are you practicing so you can make beautiful music.

**How does Epigenetics Work**

This is a complex subject that is advancing each day. It is relatively new; most of the knowledge has come about in the last 5 years. As we already stated, our DNA is divided into genes which are programs that code for a particular protein. The cell signals a gene to be turned on or expressed via epigenetics (methyl groups). The gene is copied by dividing it in half (to form an RNA strand) and sent out of the cell's nucleus to the cell itself. This half strand of DNA (RNA) is used to assemble amino acids in a particular order (dictated by the gene DNA sequence). This is a process similar to building a necklace from different colored beads. The RNA calls for a red bead, two yellow beads, and a blue bead and so on until the sequence is complete. The sequence of amino acids forms a particular protein the body needs and uses. This process is accomplished in the cell using another protein function called a Ribosome.

**Figure 4: DNA Replication**[14]

    These proteins can be used by the cell to build new cells (as in cell division), cell hardware (as that used to read the RNA and build the protein—a Ribosome), etc or it can be used as an epigenetic message to trigger other genes on or off. Think of thousands of these *"firing"* all the time. It is a very complex environment to fully understand. Scientists are contemplating an epigenome (see the advanced section of this chapter) to continue where they ended with the DNA genome. This will be far more complex and take many more years to complete. A cell can cause genes within itself or within another cell to turn on/off.
    Our cells, through epigenetics, can send signals in our bodies that will throw ON/OFF gene switches within a cell, between neighboring cells and between distant cells. Neighboring cells can communicate via direct connect (touching each other) with each other like a handshake. Nearby cells can communicate by passing signals back and forth like throwing a ball in a game of catch. Some signals, like stress, send out broadcasts like a

radio or TV signal. These broadcasts affect large number of cells throughout the body. Stress signals the body to prepare for an event. Different cells do different things to prepare. The signal stops and things return to normal when the stressful event ends. Modern life can cause us to be in a stressful state constantly. Read more about this signaling at The University of Utah's website.[15]

Our DNA does change but slowly over a very long time. It is amazing to us how similar other forms of life's DNA are to ours. The following table shows how many genes are in common with human genes.

| Type of Organism | Percent of Common DNA with Humans |
|---|---|
| Brewer's Yeast | 31% |
| Roundworm | 40% |
| Fruit Fly | 50% |
| Banana | 50% |
| House Mouse | 85% |
| Chimpanzee | 99% |
| Amount of Common DNA in all Humans | 99.9% |
| Identical Twins | 100% |

**Figure 5: Table of Common DNA**

Isn't it amazing how close we all are? Brewer's Yeast is so small we can't even see it. There are no arms or legs, no brain or head yet it shares 31% of our DNA. We consider this to be the DNA needed to control cellular functions common to all life forms. We really are not a single entity as we think of ourselves but a community of individual cells doing their own thing. They are like workers specialized in different tasks procreating via cell division a new generation of workers. They even communicate with each other for the betterment of the community. Each of us has about 10 trillion cells in our body. That is much larger than the number of humans on this planet. Each of us is a massive community of individual living cells. The community is a human being. Our community works based on what we give it. The food, the environment and how we think. What messages are you sending to your hard working cells?

In the communities we live in, we react to changes like storm damage, crime, natural disasters, etc. These outside events cause the community to react and do something about it. Some communities do things that make them better and others do things that make the community worse to live in. The cells in our body are doing the same thing. They are trying hard to adapt to changes.

## Epigenetics is NOT Hard Wired

Epigenetic controls determine who we are, what type a cell is, our health and well being. "Epigenetics tells us little things in life can have an effect of great magnitude."[16] Experiments on rats showed that young rats that were licked by their mothers, an act of nurturing, had developed bigger hippocampus (or hypothalamus) glands than rats that were not licked. The hippocampus gland is part of the brain at the top of the brain stem. It controls body temperature, thirst and hunger, as well as fatigue and sleep. It also controls our emotions. Scientists are now trying to determine if the same is true in humans. Do children that are nurtured and held at an early age fair better than those that were not?[17] The study of epigenetics may very well bridge the gap between social processes and biological ones.[18] There is very little evidence of how our emotions affect our physical being. It was not considered possible until recently. Scientists are now finding we are affected by all kinds of environmental conditions including how we feel.

## The Computer Analogy

A computer can have many programs that are never used or expressed. Think of two people that buy a computer. Identical computers with lots of identical software loaded on them. Person A uses a word processor, email and a digital picture program. Person B uses a spreadsheet, email and a video editing program. Although both machines are identical (like identical twins DNA), they have turned ON different programs that make them operate very different from each other. Identical *twins* can also be very different. One can get cancer while the other doesn't. Even though their DNA is identical, their epigenetics is quite different. It was always a mystery to scientists that identical twins with identical DNA could be so different.

A question about changing epigenetics in twins was asked of Randy Jirtle, Director of the Laboratory of Epigenetics and Imprinting at Duke University.[19]

**Q**: I don't see how twins demonstrate epigenetic changes as they differ with age. Aren't there other factors at play that have nothing to do with the epigenome?—Anonymous

**A**: If twins split from a single egg were absolutely identical, both the genome (DNA sequence) and epigenome (DNA methylation, histone marks, etc.) would be exactly the same. This is not; however, what is observed (Fraga et al., Proc. Natl. Acad. Sci. USA 102: 10604-10609, 2005). Although the epigenomes are similar in young twins, they are increasingly different in older twins, especially if the twins have lived in different environments and have had different social habits (e.g., smoking, drinking, and eating). Interestingly, identical twins can also vary in their susceptibilities to diseases and their physical and behavioral characteristics. Although there are several possible explanations for these observed differences, one is the existence of differing epigenomes.

It is amazing that as the above twins age, their epigenes differ more and more. Each is exposed to differing environments, foods, cares, and thoughts. So what influences epigenetics? It turns out almost everything! Food, environment, how we think, and music we hear things we think about, etc. A human starts with a set of DNA and epigenetic material from their parents (and even grandparents). The DNA is hard wired and not easily changeable but the epigenetics can be changed by us. As a species we can drastically change; most of us experience changes to the environment, food, how we think, etc. Look at how our food habits have drastically changed from the early 1900's. In general, the US, doesn't eat as many vegetables and fruits as we have in the past. We eat more processed foods with preservatives and bigger portion sizes. We also are experiencing much more chronic illnesses. Diabetes, heart problems, obesity and cancer have become the norm. Can we undo what we have done and get back to a healthier life both physically and mentally? The answer is YES! This book is about the problems and the potential solutions.

## Diseases can be Switched on by Epigenetics

Professor Randy Jirtle, Duke University Epigenetics department studied both mice and human identical twins and found:[20]
- Identical twins do not always get same diseases
- He was able to turn on a gene in identical twin mice making one twin fat and the other thin

- You are what you eat but also what your mother ate and maybe what your grandmother ate
- A study was done of groups of identical twins from age 3 to 74. The genes and epigenes of each set of twins were overlapped. Younger twins had a lot of similar epigenes but as twins aged they had less similar epigenes. What they ate, if they smoked influenced their epigenes making them different from each other.

Identical twins, essentially, start life with identical DNA and epigenetics. As the egg splits into two, the similarity stops! The DNA will continue to be identical but each fertilized egg will react differently to its environment. These changes will continue to differ as it grows into a fetus. Finally as they are born and mature as adults the changes in their epigenome will be more pronounced.

Professor Jirtle says: "The good news is fixing the epigenome is a lot easier than fixing our genes" but "it is a lot easier to mess it up as well". "We all have a responsibility and a hope for our epigenes." Dr. Bruce Lipton, in his "The Biology of Belief: An Epigenetic Primer", describes these communities of cells as:[21]

> Even though humans are made up of trillions of cells, I stressed that there is not one "new" function in our bodies that is not already expressed in the single cell. Each eukaryote (nucleus-containing cell) possesses the functional equivalent of our nervous system, endocrine system, muscle and skeletal systems, circulatory system, integument (skin), reproductive system and even a primitive immune system, which utilizes a family of antibody-like "ubiquitin" proteins.
>
> I also made it clear to my students that each cell is an intelligent being that can survive on its own, as scientists demonstrate when they remove individual cells from the body and grow them in a culture. As I knew intuitively when I was a child, these smart cells are imbued with intent and purpose; they actively seek environments that support their survival while simultaneously avoiding toxic or hostile ones. Like humans, single cells analyze thousands of stimuli from the microenvironment they inhabit. Through the analysis of this data, cells select appropriate behavioral responses to ensure their survival.

If we look at it from a DNA point of view rather than a human one; different humans react differently as their epigenetics are different and different genes are expressed and suppressed. Some human's will die early because bad decisions were made at the cell level while others will survive. Those that survive will have epigenetics and genetics that work well in a given environment. The species moves on and so does DNA. Are we then slaves to our cells and their DNA? No, we can send different signals to our cells that will make us happier and healthier. In other words, our cells don't always make decisions that are best for the community (body) because we don't give the cells what they need.

**Epigenetic Inheritance**

Nova ran a segment on epigenetics and inheritance that: [22]
- Introduces epigenetics and notes a key difference between genetics and epigenetics. Genetics is the study of DNA-based inherited characteristics in organisms while epigenetics can involve modifications to DNA (i.e., DNA methylation) or changes to the structures surrounding the DNA (i.e., the chromatin's histones).
- Details one epigenetic mechanism by which methyl groups attach to the histones, affecting the expression of the DNA's genetic code.
- Explains that chemicals that may have an epigenetic effect, such as methyl groups, can enter the body from one's environment. Consequently, an individual's choices and personal history, such as diet and smoking, can influence his or her exposure to epigenetic triggers and their accumulation in the body.
- Describes studies that demonstrate the effect of epigenetic factors. For example, one set of genetically identical mice was fed food rich in methyl groups. The methyl groups bonded to the mice's chromatin and blocked the expression of certain genes. This epigenetic effect produced two sets of mice that looked extremely different.
- Reports on studies of identical (human) twins that reveal a measurable accumulation of epigenetic changes as the twin's age.
- Notes the development of a promising line of therapies that work by rearranging a cell's epigenetic tags. One epigenetic therapy mentioned attempts to alter the behavior of cancer cells.

## Epigenome

Obviously knowing what environmental changes trigger which epigenetic reactions would help us become a healthier species. Biologists are starting a new genome called the epigenome. It is designed to answer these questions. We have far more epigenetic reactions than genes so it will take some time to complete this study. Certainly it will take more than our lifetime. Is there anything we can do now? Can we make ourselves healthier, happier, and calmer? These are the questions the book is designed to help you answer.

## *Science*

### Methyl Groups in Foods

Our DNA is organized into genes which are tightly wrapped around a protein called a histone. Epigenetic markers cause these histones to bind tightly together or fall apart loosely. When bound tightly the genes inside cannot be expressed and therefore are OFF. When loosely bound they can be expressed. This is one form of epigenetic control.[23]

The way epigenetics works from a biological point of view, is through "methylation". Methyl groups cause epigenetic triggers. Which foods contain methyl groups? What are methyl groups? A "methyl" group is simply one carbon connected to three hydrogen atoms. It may be written as $CH_3$." Methylation" is not just one specific reaction. There are hundreds of "methylation" reactions in the body. Methylation is simply the adding or removal of the methyl group to a compound or other element. One form of methylation tightly binds the genes to a histone protein so they cannot be opened and read (disables the gene). Loose bindings can be open by the nucleus' mechanisms and transcribed into RNA (epigenetic expressed). The RNA then travels out of the nucleus into the cell's main part to be used to build a particular type of protein.

So why do we care about methylation at all? In general, when some compounds receive a methyl group, this "starts" a reaction (such as turning a gene on or activating an enzyme). When the methyl group is "lost" or removed, the reaction stops (or a gene is turned off or the enzyme is deactivated). Some of the more relevant methylation reactions would be: [24]

1. Methyl groups "turns on" detox reactions that detox the body of chemicals, including phenols. So if you are phenol sensitive, and

you increase your methylation, then theoretically your body can process more phenols and you can eat fruits without enzymes!
2. Getting methyl groups to "turns on" serotonin, and thus melatonin, production. Therefore, if you are an under-methylator, you can increase your methylation and have higher more appropriate levels of serotonin and melatonin. This means you may not have to take SSRIs, or may have improved sleep.
3. If you are an over-methylator you can take certain supplements to decrease methylation and perhaps turn off reactions that need to be off. This may decrease aggression or hyperness, for example.

It is clear that foods affect methylation and methylation affects epigenetics by turning ON/OFF our genes. A side step that plays an important part of our health is that they also turn ON detox reactions. These allow our bodies to better fight disease. It is also clear that getting the right amount of something can be beneficial but too little or much can cause problems. It's a delicate balance but it was developed over millions of years as we evolved. We don't need to study methylation groups all we need to do is eat as humans have for so long.

We have all been hearing about the positive effects of Omega-3 fats. Fat is a molecule that is made of a fatty acid that contains on one end an acid group (-COOH) and the other end contains a methyl (-CH3) group. Methyls again! They are in our foods and they affect our genetics.[25]

**Epigenetics in Science**

Epigenetics has been linked to:[26]
- Fertility
- Brain Tumors
- Immune System disorders
- Breast Cancer
- Colon Cancer
- Parkinson's
- Skin Cancer
- Prostate Cancer
- Leukemia

These are current areas of study involving epigenetics. We probably will find that epigenetics is responsible for most of our health problems.

If this turns out to be true, it is not epigenetics but what environmental change caused it to change our epigenome.

**Cancer Research**

When researchers compared normal and malignant cells, they found the cancer-fighting gene, P16, had been switched off in the malignant cells through a process called DNA methylation, where molecules are added to the DNA backbone affecting its function.[27] Researchers are looking to use this technology to develop drugs that turn OFF or ON particular genes. We better be very sure we understand the epigenetic genome before we start changing it. A better way would be to study what external factors in the body cause these epigenetic switches in the first place. If it is food we can change our diet. If it is stress we can learn yoga and meditation. If it is thinking we can learn to think differently. We have choices. Let's study all possible paths so we truly understand what is happening and how to fix the original cause not just the symptom.

In another scientific study researchers discover role for epigenetics in cellular energy regulation.

> A breakthrough on how cells regulate their energy is promising for clinical gains into diseases such as obesity, diabetes and cancer. Researchers at McGill University and University of Pennsylvania have uncovered new insights into how a protein known as the AMP-activated protein kinase, or AMPK, a master regulator of metabolism, controls how our cells generate energy. AMPK has previously been linked to a number of biological functions including cancer, diabetes, and proper immune function. AMPK mediates "epigenetic regulation" of gene expression by binding directly to sites on chromosomes.[28]

## *Conclusion*

**The More we Know . . . The Less we Know**

Ten years ago we all thought we were somewhat predisposed to our genes. Some of us got *good* genes and others *bad* genes. The media started telling us there was nothing we could do. Our intelligence was predisposed; our health was predisposed, etc. They couldn't have been more wrong. Epigenetics shows us that it is not what genes we have but how they are played that counts. We do inherit our genes from our parents and we also inherit our epigenetics from our parents and probably our grandparents as well. So what is the difference . . . genetics or epigenetics? Genetics (our DNA) is not easily changed. It takes many generations to have small changes in our DNA. Radiation can change them but in random ways. Epigenetics on the other hand can be changed by us. We simply need to change our food, thought and environment. It may not be very easy to do this but we can do it.

## Understanding our Epigenetics

Scientists want to map the epigenome. This task is much more difficult to understand. We must map which amino acid chains and protein chains activate or deactivate which genes. We also must understand where these activators and deactivators come from. Are they manufactured by other cells in our body (like our brains)? Are they consumed in our food supply? Are they produced by our brain by the way we think? Are they ingested from our environment? Maybe several sources are possible. Will we then really understand everything about human life or will we discover that God and nature are even more complex than we ever imagined.

Physicists go through this all the time. The search for the basic building blocks of life has been going on since Plato. The Greeks believed matter was made up of tiny hard particles like a small ball they called atoms. We then discovered that the atom was mostly air made up of a very small nucleus and one or more electrons circling the nucleus. Later we found out the electrons were made up of even smaller particles and so was the nucleus. These particles were made up of even smaller ones. The latest theory, string theory, is that the smallest elements are called strings of energy that vibrate at different frequencies (like music). Each frequency creates a different subatomic particle. We believe the study of epigenetics will reveal more questions than it answers. Ultimately we will again be reminded that God's creations are infinitely more complex than we imagined.

Dr. Bruce Lipton speaks about "How epigenetics works":[29]
- Scientists have looked but cannot find a specific cancer gene. Instead they are finding that the 200 different genes in a cancer cell have different epigenetic markers than do normal cells.
- Your beliefs can change your genetic expression. How you think can cause different epigenetic results. This is why the placebo effect is real. We believe it will happen and it does.

- There are spontaneous remission cases. People with cancer or other deadly illnesses are given a short time to live with no hope for a cure but then our bodies cure themselves spontaneously.
- People around us control our perceptions. These include our family and community.
- Perception controls epigenetics and which genes get expressed or blocked.
- Philosophy, religion, culture is important because it controls our perception of the truth and therefore our epigenetics
- Jesus said: "You can renew your life in your beliefs". He knew the power of perception over our genetics.
- Your belief can heal you or cause you to be sick and die. It is only a matter of positive vs. negative beliefs. Both are equally powerful.
- Whether you think you can or think you can't, your right either way said Henry Ford
- Placebo is a positive thought having positive actions like healing you.
- Nocebo is a negative thought having negative actions like making you ill or dying.
- We have all been programmed in our society to be a victim. Therefore we are victims.
- Self healing works! Why can't we use it better? Most of us have bought into the idea of our genes control us. Some of us have good ones and others bad ones. Modern biology shows us this is totally false. Our genes do nothing until they are turned on (expressed). Epigenetic factors express our genes.

The primitive notion that our DNA is our destiny is giving way to the understanding that our genes do nothing until they are activated. Environmental conditions (including not only the chemicals that enter our body but also the decisions we make, the people we hang with, and the stress we undergo) determine whether a gene gets turned on or off. Our genetic array is like a keyboard, and our interaction with the world governs what melody gets played on it. [30]

Epigenetics is a branch of biology that is moving at light speed. Each day you can find new papers and references to it on the internet. We are only in our infancy of understanding how it really works. Epigenetics will become a household term just like DNA did.

Our biology is certainly complex. Who will determine what affects epigenetics? In the next chapter we will see that another relatively new area of biology is nutrigenomics or the study of which epigenetic triggers are set off by which foods.

**Largest ever Epigenetics Project Launched**

On September 8, 2010, one of the most ambitious large-scale projects in Human Genetics has been launched: Epitwin will capture the subtle epigenetic signatures that mark the differences between 5,000 twins on a scale and depth never before attempted, providing key therapeutic targets for the development of drug treatments. The project is collaboration between TwinsUK, a leading twin research group based at King's College and BGI. [31]

If you are a young student looking for a fun and rewarding career, look into biology: epigenetics. This field will become a household term and deliver earth shattering results over the next several decades. If you are an investor, look into companies building new epigenetic related products.

# Chapter 3:

# Nutrigenomics

## *Introduction*

Nutrigenomics is the study of food nutrients and their effect on disease. This field is very young still but is exploding! A Google search produced over 1,040,000 hits for Nutrigenomics in June of 2011. This is a critical field of science that will create links between the food we eat and the diseases we contract. Our diet affects epigenetics which can affect our health.

For those of you reading this, who are investors, the areas of epigenetics and nutrigenomics will be great investment areas for years to come.

## *Basics*

### Disease and Nutrigenomics

Cancers have been observed to have high levels of methylation (epigenetics). We now know that foods as well can aid in the prevention of cancer. Is there a link between certain foods and cancers?[32] These studies suggest that there are GOOD and BAD foods. Bad ones include: Red Meat, processed meat, grilled meat, dairy, animal fat, partially hydrogenated fats, etc. Good foods include: fish, fruits, vegetables, tree nuts, omega-3 fatty acids, whole grains, etc.

Cancer, once believed to be caused by mutations of our genes, is now believed to be caused by mutations that are turned on via epigenetics. [33]

Epigenetics is one of the most scientifically important, and legally and ethically significant, cutting-edge subjects of scientific discovery. Epigenetics link environmental and genetic influences on the traits and characteristics of an individual, and new discoveries reveal that a large range of environmental, dietary, behavioral, and medical experiences can significantly affect the future development and health of an individual and their offspring.[34]

## It starts in the Womb

The research, presented at the American Association for Cancer Research's annual conference, is one of several to show that conditions in the womb can affect health for generations. The researchers, from Georgetown Lombardi Comprehensive Cancer Center in Washington DC, then tried to work out how something that happened in pregnancy can go on to affect the health for generations to come.[35] Mothers-to-be who gorge on junk food could be putting their grandchildren at risk of breast cancer.

> They showed that it wasn't due to the junk food diet raising levels of estrogen, a hormone that fuels the growth of breast tumors. Instead, they believe it can be explained by a process called epigenetics, in which conditions in the womb cause subtle changes to the way genes work. These changes, which are different to mutations, can be passed down the line from mother to daughter or from father to son, time and time again. In this case, the tiny changes may increase the number of potentially cancerous 'buds' in the breast.

Our lifestyle choices don't just mark siblings in a sociological way but also may be marking their genomes with chemical tags that tell some genes to turn on and others to stay off. These effects on gene expression have been seen to have both long-term and wide-ranging effects on health. They might explain the correlations that researchers have found between lifestyle and risk of disease.[36]

> At one point we thought our genes were static, meaning that if your mom had arthritis . . . you were very likely to get it too. While this is still true, we now know that what we eat, how

we think and move can actually influence our genes. The good news . . . we can talk to our genes and influence which of these disease factors will be expressed.[37]

Using the computer analogy again, we can pass on a computer with hardware and a set of programs to someone else. It will be hard for them to change the hardware (without buying a new computer) but relatively easy to change the software. If they don't change the software they will be stuck with the same programs, bugs and outputs that the machines donor had.

## *Advanced*

Science is quickly creating research groups that will investigate the areas of epigenetics and nutrigenomics. Every major university has changed their biology curriculum to include these new topics. The textbooks that exist are all old and outdated! New textbooks must be written to include research in these areas. Over the next few years we will see an explosion of new nutrigenomics that will affect who we are and how we eat.

### The Nutrigenomics Organization (NuGO)

NuGO is an Association of many Universities and Research Institutes focusing on jointly developing the emerging research area of nutrigenomics and nutritional systems biology. NuGO evolved from a European Union (EU) Sixth Framework Network of Excellence, and is now expanding to global dimensions. You can join via an Institutional membership

**NuGO has two major objectives**:
- Stimulating developments in nutrigenomics, nutrigenetics and nutritional systems biology, and
- Incorporating these aspects in nutrition and health research, by joint research projects, conferences, workshops and training. The NuGO can join as partner in your research project, being a legal entity.

They will be shaping the **nutrition bioinformatics infrastructure,** by initiating, coordinating, facilitating projects in this area and by hosting and disseminating all data, results and information in this area.

# *Science*

## Prevention is Better than a Cure

You can change your genetic destiny by changing the way your genes are expressed, much in the same way you can make music louder by turning up the volume by pressing the volume control button.[38]

The science of individuality exposes those genetic traits that can be naturally awakened or suppressed, often through the foods we eat. Being attuned to your particular, unique physiology—the gene mix you have inherited from your ancestors means:
- you can significantly reduce your health risks
- optimize your weight
- maximize your physical and intellectual performance
- increase your overall quality of life
- lengthen your life span

Nutrigenomics deals with how diet interacts with one's genetic make-up to affect one's health. It is the study of how different foods interact with particular genes, affecting how these genes act or altering their structures. Specifically, nutrigenomics is concerned with how chemicals in different foods can interact with particular genes to increase the risk of diseases such as type 2 diabetes, obesity, heart disease, and some cancers.

Nutrigenomics dictates that understanding how dietary chemicals regulate different genes will lead to individualized nutrition, the ability to design diets catered to one's specific genetic make-up. For example, the food pyramid developed by the USDA assumes that all Americans are the same and have the same dietary needs. Of course, the truth is that we're not.[39]

# *Conclusion*

Both epigenetics and nutrigenomics are fields to follow over the next 20 years. At the conclusion of the epigenome study we will find that the food related data (Nutrigenomics) is what really matters for diet. Today we are changing our diet drastically and how the food is prepared.

In the past our cells would get changes slowly over time or one big change to deal with. Environment sometimes changed quickly (like after a major meteor hit Earth) or slowly over time. While this was happening diet mostly stayed the same. Today we are changing it all at once. Can our cells modify their genes via epigenetics fast enough to save us? We hope so. Both epigenetics and nutrigenomics will be at the forefront of answering that question.

What foods cause bad epigenetic reactions? What cause diseases? How do stress and our belief system affect our epigenetics? These questions will be answered in the next few chapters as we look at the parts of our environment that affect epigenetics. We will not be covering air quality or chemicals we use and how they affect us in this book. Most people understand that if you are around chemicals you can get sick. This book will deal with three parts of the environment:

1. Food
2. Thought
3. Belief

# Chapter 4:

# Environment—You are what you Eat, Smoke and Drink

### *Introduction*

Does something as basic as eating cause epigenetic changes that can affect our health and well being? The answer seems to be YES! Food provides us with the missing nine essential amino acids that our body cannot reproduce on its own. Eating also provides essential things such as fat, carbohydrates, minerals, etc. that our bodies also need. Food also contains methyl groups (chemicals) that are some of the basis for how epigenetics work. These groups have the ability to switch ON/OFF genes. In this chapter we are going to look at some foods that can cause us problems, foods that are healthy for us and try to correct some old myths.

We all eat for a few reasons:
1. We get hungry—hunger is a pain that gets stronger so it motivates us to eat
2. We get pleasure from eating certain foods—Sometimes these pleasures are for nutritious foods and sometimes for non nutritious foods

Are you eating to live or living to eat?

## Basics

### Our Food

Studies have shown that proteins, methyl and chemicals in our food can cause epigenetic triggering that turn genes on or off causing all kinds of disease or preventing diseases depending on the food. The National Academies tackled this topic stating chemicals can turn genes on and off but new tests are needed.[40] Food is an important part of being healthy. People have always known this but epigenetics is now showing us that this is even more far-reaching than originally thought.

In Chapter 1: Early Life, Early Man and DNA, we saw that proteins can interact with the cell membrane through portals. This causes reactions inside the cell. Other proteins gather inside the cell at the portal in response to the external proteins present. These internal proteins travel inside the nucleus and express (turn on) a particular gene. This action is epigenetics. So external proteins in our bodies can cause genes to express or not express based on the proteins make up (amino acids) and the shape of its folding.[41] Proteins are a string of different combinations of amino acids that fold into patterns. Both the combination of amino acids and the folding are important to various body functions. Sickle cell anemia is a disease caused by a miss-folding of a protein strand in the blood. The miss-folding causes the platelets to not be able to carry an oxygen molecule.

Our diet and our levels of stress can cause these reactions. Our bodies are a multitude of living organized and intelligent cells working together and communicating with each other to determine what is best for the organism and the continuance of the DNA code.[42] Even Plato recognized the importance of diet and stress.

> *"And we have made of ourselves living cesspools, and driven doctors to invent names for our disease"*—Plato

From early man to present man, life didn't change much from a DNA point of view and even the epigenetics would be similar until very recent times. What changed all of a sudden and had a profound impact on our lives? Food! We started to mass produce it and reduce its nutritional value to basically nothing. Processed white bread, cookies, candies, cakes, canned goods, boxed prepared foods, etc are all relatively modern additions to our diets as is fried foods. Most modern people eat three meals a day which are way too large and the body gets too many extra calories. Sometimes modern people will skip a meal or two thinking it will make up for last night's big dinner. Since we are eating one, two or three big meals instead of 5 or 6 small ones, the "Starvation Mode" mechanism cuts in and causes the extra calories (and there are a lot of them) to be stored as fat.

This adversely affects our health. Early man was very healthy with eating many small meals and getting plenty of exercise. Modern man has little or no exercise and is not eating healthy at all. We currently see advertisements on TV that show someone not wanting to eat their vegetables and being shown we can get them from a juice or in a product that doesn't seem to contain any vegetables except maybe some tomato sauce and much less fiber than the fruit or vegetable would contain in its natural state. At the same time America has become the sickest nation on Earth. With most people considering it normal to get high blood pressure, cholesterol, heart attacks or cancer. These illnesses are NOT in our DNA or epigenetics unless we put them there. By eating a bad diet, we modify our epigenetics causing some DNA (genes) to activate and others to deactivate. The net outcome is that we are in much worse shape than early man was. His only concern was predators but he lived a healthy life while alive.

## What has Changed in our Diets in last 11,000 years

Dr. Andrew Weil, MD sites a study, that was done by a researcher at the Harvard School of Public Health, in his book Eating Well for Optimal Health[43]. The study was done to list the top 20 American foods that are consumed. Here they are:

1. Potatoes*
2. White Bread
3. Cold Cereal
4. Dark Bread
5. Orange Juice
6. Banana*
7. White Rice
8. Pizza
9. Pasta
10. Muffins
11. Fruit Punch
12. Coca-Cola
13. Apple*

14. Skim Milk
15. Pancakes
16. Table Sugar
17. Jam
18. Cranberry Juice
19. French Fries
20. Candy

*These 3 foods were around 11,000 years ago (potatoes and bananas were found only in a few areas), so that is 3 out of the top 20 eaten today. Most potatoes today are consumed as French Fries. Beginning to see how we have changed? Many people have said to me "We all must die sometime" This is surely true but how we live our last ten or twenty years is under our control. We can be sick and immobile, a burden to our friends and relatives or we can be healthy and enjoy life up to the end. It's your choice, your free will. Choose wisely! We have seen many articles on elderly men and woman in their 70's, 80's, 90's and even 100 that are competing in sporting events and winning. Illness is not a verdict of age.

**You are what You eat**

The saying "You are what you eat" has been around for a long time. People usually believe this at some level. Most people don't realize how bad not eating right can be. In Chapter 2: Epigenetics, we discussed what epigenetics is and how it can affect our quality of life. We derive vitamins and minerals from food but also sugars (both good and bad) and protein (not just animal products). We know that protein is the bases of our cellular activities. The cells of every living thing on earth are made of protein. Therefore fruit, vegetables as well as fish, poultry and meat provide us with protein. Earlier we discussed how the cell as a factory needs certain raw materials to operate efficiently.

We find that when we eat healthy with lots of fruits and vegetables, our palate enjoys these flavors and wants more of them. When we eat fried foods or sugary foods our palates want them. Train your palate to want and enjoy healthy foods by eating more of them. During the summer months find locally grown produce and eat a lot of it. The flavors will amaze you and awaken your taste buds. Your weight will normalize to a more healthy level and your spirit will feel happier about who you are. Food is so powerful in our lives.

Does anyone reading this really believe that man can manufacture better foods than God or nature? Eat "God created" carbohydrates not man's processed versions. Your body and genes will react more favorably to the natural versions. Try to find organically grown ones to minimize toxins to your body. Try to buy low sugar and salt versions to reduce hidden salts and sugars. Buy locally grown produce since it will be fresher and tastier. Fruits and vegetables were designed to be eaten from the plant. The longer you wait from picking them to eating them, the more nutrients they will lose. We can all start cooking more from raw materials and buying less prepared foods that we microwave. Go through your pantry and look at the food supply you currently have. How much is in a box, can or bag? How many have preservatives, salts, sugars in it? Is the main ingredient first on the list of ingredients? How much fiber is in the product? If not what are you eating? Eating nature's foods is easy; we know it has the ingredients we need to be healthy. We know it because these foods evolved along with us.

David Shenk says, "Australian geneticists Daniel Morgan and Emma Whitelaw discovered that mouse fur color can be manipulated by something as basic as food."[44] This is absolutely amazing. Food can affect something as basic as fur color. It would appear we really are what we eat. These types of changes were thought to be in the realm of our DNA only. We now know that DNA is only part of the equation. The other part, the important part is how that DNA is played or expressed. If the color of a mouse's fur can be changed by food, think of how your health can be impacted by food. We truly are what we eat!

We are given guidelines on what to eat, for example, the food pyramid. Public school nutritional guidelines from 1915 to present have been provided by the National Dairy Council.[45] The US Government shows the percentage of protein in different fruits and vegetables. You may be surprised to know that Spinach is 49% protein. Beef is only 27%. In fact, most green vegetables are 20—50% protein and algae's are up to 70% protein.[46] Why didn't we learn this in school? The answer is the National Dairy Council does not sell spinach or green vegetables. Never be afraid of where you will get your protein. It is in every food you eat. Do we get enough protein? Most Americans eat far too much protein. Excess protein is toxic and over time, can cause liver problems. Chapter 14: How do you get your protein, covers this in detail.

> *The body is not a frozen structure.*
> *It is a river of information—*
> *A flowing organism empowered*
> *By millions of years of intelligence.*
> *Every second of our existence*
> *We are creating* a *new body*[47]

Most of us think of ourselves as alive from birth to death. We are an entity growing older each day. In reality we are made of a community of living cells which are constantly dying and being reborn. Our bodies today are a clone of what they were yesterday. We are never the same person but a person made up of a community of cells. These cells are changing all the time. Each day of our lives we need to provide the proper nourishment for our cells. Just as we provide proper fuel and oil and water for our automobiles. If we neglect this for too long, it will fail us. We should be consciously making the choice of what to eat before putting ANYTHING into our mouths. For most of us, that is not what is happening.

The food industry is a very powerful business that, like all businesses, wants to make as much money as possible. They are very good at giving us the junk foods we seem to want. Sugars are often added to make the foods more appealing and addictive. Since they are mass produced, preservatives must also be added to increase the shelf life and profitability of the product. We seem to desire things that are full of fats, sugars and salts. This can be broken by eating a healthy diet of fresh whole foods. It is amazing how our taste buds adopt to what it knew for millions of years. These three substances (fats, sugars and salts) do trigger a pleasure response from our brains, making us want them more.

The food industry is also very good at brain-washing us with information it wants us to believe. Commercials for these foods like one saying high fructose is OK for you in small quantities play on our feelings that anything is OK in moderation. The problem is we don't end up eating them in moderation and the industry knows that. Another commercial shows adults not wanting to eat their vegetables, feeding them to the dog and getting their fruit and vegetable servings from a bottle of juice. Do we really believe deep down, that nature's natural foods that we have been eating from the beginning of time are not as good for us as man's bottled and canned products that are 20 or 30 years old or less? These foods are toxic to our body causing our liver to work hard to rid us of their affects. Read the list of ingredients and see what you are really eating.

Eating the wrong foods actually starve our bodies of nutrition while bloating it with calories. Most overweight people are starving themselves of essential nutrients. This type of eating leads to binging and desiring food that is not nourishing. We eat more and more, gain weight and starve our cells of what they need to be healthy. Think about what you eat, read labels, avoid processed foods and return to a healthy diet.

Dr. Sing in his book, "Food and Medicine" shows this cycle, and the way it evolved, was documented a year ago by David Kessler, a former commissioner of the Food and Drug Administration and former dean of Yale's medical school, in his ground-breaking book "The End of Overeating"[48].

**Food labels**

Food labels can be confusing. I have seen instant mashed potato mixes that list potato way down the list of ingredients. These lists start with the item mostly contained in the product and proceed down to the item least found in the product. If mashed potatoes aren't using potatoes, what are we eating? When reading a label, look at the ingredients to see what is in the product. Are there preservatives? If you can't read an item, YOU SHOULDN'T PUT IT IN YOUR MOUTH. Next, look at fats, salt and sugars. Are they high? Are there saturated or Trans Fats? Finally look at the fiber. You want fiber in your diet. Below are two labels reproduced as examples:

| Amounts per Serving | As Packaged | As Prepared |
| --- | --- | --- |
| **Calories** | 240 | 280 |
| Calories from Fat | 25 | 60 |
| **Total Fat 3g** | 5% | 11% |
| Saturated Fat 1.5g | 7% | 21% |
| Trans fat 0g | | |
| **Cholesterol 5mg** | 2% | 6% |
| **Sodium 690mg** | 29% | 31% |
| **Total Carbs 49g** | 16% | 16% |

| Dietary Fiber 3g | 13% | 13% |
|---|---|---|
| Sugars 2g | | |
| **Protein 7g** | | |

**Figure 6: Box of Rice Label**

We would typically cook whole rice from scratch. Using a boxed version sounds quicker but it is really not. You get extra calories, sugars, salts and maybe preservatives that you don't need. This label has a big red flag for us since it contains 31% of your recommended daily salt intake. Eating a double portion provides us with 62% of our recommended daily intake. This is way too high. Be very careful of products, especially soups with high salt contents.

| **Amounts per Serving** | **% Daily Value** |
|---|---|
| **Total Fat 1g** | 2% |
| Saturated Fats 0g | 0% |
| Trans Fats 0g | 0% |
| **Cholesterol 0g** | 0% |
| **Sodium 820mg** | 34% |
| **Total Carbs 12g** | 4% |
| Dietary Fiber 7g | 28% |
| **Sugars 0g** | |
| **Protein 11g** | |

**Figure 7: Jar of Beans Label**

Beans are very good for you. If you wash them well before cooking them, they will not give you gas. The label above shows high sodium again. Beware that sodium is used as a preservative. You get much more sodium from prepared foods than from your salt shaker. Also note that beans have high fiber content. Fiber is essential in making our bodies function properly.

## Food Additives

We have been speaking about natural foods throughout this book. Obviously eating things with artificial additives and preservatives can be bad to our bodies. It goes without saying that these additives can cause health issues as well. The question is even with all the additives can we cure ourselves because we do eat plant based foods and believe we can? The power of our minds is only partially understood. We now know there is a link between our mind and our cells. This link can cause thinks to happen at the cellular level that may be good or bad for us as an organism. George makes homemade wine. Many people have said to him that they can't drink red wine because it gives them headaches or acid reflux. These same people drink his red wine and never get headaches or acid reflux. He uses very little preservatives (about the same as nature uses) and makes the wine as naturally as possible. What we ingest does matter! What kind of fuel you put in your car determines how it will perform and what you put in your bodies determine how they will perform.

## *Advanced*

### *Food Quality*

Studies have shown eating less overall calories, but not going into starvation mode, can extend our life. Dr Weindruch did a study on rats to determine the affect of larger or smaller meals. The rats that could eat all they wanted experienced more stress and less overall energy while the rats that ate less experienced just the opposite effect. UCLA has done similar studies on larger animals such as monkeys and even humans. The results were similar[49]. Others countries eat smaller portions as the norm. They experience less obesity than the US does. Asian countries have people that are very thin and healthy but after coming to America and beginning to eat like Americans, they gain weight and start to get the same illnesses that plague Americans.

We Americans have a tendency to think of other countries like China as backward and poor. They eat rice and fish, and we eat steaks. In reality they are stronger and healthier than we are. They have far fewer cases of chronic illness and are far less obese than we are. So who is better off? The problem is we are moving into a global economy and our American foods and fast food chains are being moved out to the world. As we do this the world is starting to become more ill and more obese. What a legacy.

Correct portion sizes are much smaller than we Americans are accustomed to. Here is a good way to visualize a correct portion size:[50]
- Woman's fist or baseball—a serving of vegetables or fruit is about the size of your fist

- A rounded handful—about one half cup cooked or raw veggies or cut fruit, a piece of fruit, or ½ cup of cooked rice or pasta—this is a good measure for a snack serving, such as chips or pretzels
- Deck of cards—a serving of meat, fish or poultry or the palm of your hand (don't count your fingers!)—for example, one chicken breast, ¼ pound hamburger patty or a medium pork chop
- Golf ball or large egg—one quarter cup of dried fruit or nuts
- Tennis ball—about one half cup of ice cream
- Computer mouse—about the size of a small baked potato
- Compact disc—about the size of one serving of pancake or small waffle
- Thumb tip—about one teaspoon of peanut butter
- Six dice—a serving of cheese
- Check book—a serving of fish (approximately 3 oz.)
- Eyeball it! Take a look at the recommended serving sizes on the new USDA MyPyramid Food Guidance System. Get out a measuring cup or a food scale and practice measuring some of your favorite foods onto a plate, so that you can see how much (or how little!) a ½ cup or 3-ounce serving is. This will help you "eyeball" a reasonable serving!

We eat foods because we think they are healthy. The Food Pyramid lists foods in proportions we should eat but it is produced by the Dairy Association to sell dairy products. Don't be fooled by clever marketing pitches. Ancient man ate mostly plant-based foods, maybe a little meat when he could find it. Modern man eats lots of meat, very little plant-based foods, a lot of salt, sugar and fatty foods. It is no wonder our cells are rebelling and making us sick.

## America Got Fat

We were not always a nation of fat people. We did not have terms like "Super Size Me." Boy did that ever super size us in general. Our portions were not always so out of whack with the rest of the world. We did not always have such an abundance of artificial foods at our disposal. We did not always think cooking was a chore and took too long so let's go out for fast food. We did not always have the amount of sick individuals as we do today! David Kessler, in his book "The End of Overeating" dedicates his book to understanding what happened."[51] He talks about theories of treating this disorder and food rehab. This is an excellent book to better understand the industry and how it has targeted us.

I have friends from Italy and when they first came to visit America were shocked at our portions. It soon became a joke that everything in America is Texas (super) sized. We got our wish, our body mass is now Texas sized. We are one of the few countries that have restaurants that offer a free meal if you can eat the 32oz steak. Why? Food is not a game. It is our support to a healthy lifestyle.

In 2007, over 30% of the population in 3 states was obese. In 2008 it was over 30% in 9 states. This represents 2.4 million more people or a total of 72.5 million people are obese in America.[52] The cost of this problem is a whopping $147 billion/year. This condition is linked to heart disease, cancer, diabetes and stroke. The reasons given in the NY Times article was:
1. Too little exercise
2. Too little fruits & vegetables
3. Too much sugar, fat, French Fries, soda and sugary drinks

## *Protein*

Protein makes up most of the weight of a human body that is not water. Protein is essential to our bodies and to the cells of all living things. The cell walls are made of protein. Protein also makes the various cell mechanisms (ribosome) that read RNA, assembles amino acids into proteins, etc. All protein is made up of amino acids. Humans have 20 different types of amino acids that make up over 100,000 different proteins.[53] Our body has many amino acids at its disposal to do this job. It can manufacture all but nine of them. These nine are known as the essential amino acids because they have to come from outside the body in the form of food. All life has protein made up of amino acids. Fruits, vegetables, fish, fowl and meats all contain protein. We have been taught to think of protein only in terms of meat but this is not true. Scientists discovered that meats have all nine essential amino acids in their proteins. Meats have all nine because the animal eats a variety plants or other animals that eat plants. These amino acids basically come from plant life. If we eat a variety of plants we will get all nine amino acids. Once again marketing has played a role in making protein synonymous with meats. Are you starting to see how ads and commercials affect how we think and what we believe? Think about a cow! It provides us with meat and protein but what does a cow eat? It is vegetarian. Somehow it produces all nine amino acids from vegetables. So do we! Don't be fooled by advertisements intended to get your hard earned dollars. Eat healthy and live longer.

Dr Andrew Weil, MD, says that when a body gets more protein than it needs, it may increase the workload of the digestive system and place a particular burden on the liver and kidneys[54]. Dr Weil also recommends 10%-20% of your daily calories to be protein. If you have a daily intake of 2,000 calories that is 200-400 calories from protein (all sources not just meat). This is 50-100 grams of protein.

Note a 10 oz steak yields about 90 grams of protein. But you eat fruits, vegetables, milk, eggs, etc which all have protein as well. Are you beginning to see how we overeat? We are taxing our kidney and liver with toxins from excess proteins. A large apple on the other hand is about 1 gram of protein. We could eat 50 to 100 of them. One cup of Broccoli is about 3 grams of protein. A good website to determine food calories and amounts is **http://www.my—calorie-counter.com/**. Eating a variety of fruits, vegetables and nuts will give you all the protein that you need.

## *Salt*

Salt is made of Chloride and Sodium. All animals need a small amount of salt but in large doses it can be harmful or fatal. We need Chloride for our metabolism (the process that converts food to energy). America gets most of its salt from ingredients in processed foods not the salt shaker. Yes you can use the shaker too much but most people find a little salt brings out flavors. The salts in processed foods don't taste salty so we don't even know it's there. The more processed foods we eat the more salt we ingest. Remember salt in processed foods is a preservative. Using unprocessed foods is nature's way of getting us the nutrients we need to be healthy and strong.

## *Sugar*

Sugar makes our brains happy. Scientists disagree as to the addiction powers of sugar. We believe we will learn more about addiction and our brains in the future and reclassify what is addictive and what is not. Sugar is being added to many products because it addicts us to the product. Added sugars are also unhealthy. A report in NPR sited that added sugars increase risk of heart disease.[55] [56]

We all know bad foods for kids and adults such as:
1. Candy
2. Donuts
3. Cakes
4. French Fries
5. Fast Foods
6. Juice

Sugars in fruits and vegetables are good for us. They are packaged with natural fiber that delivers them to our cells at a slower rate. Processed sugars and juices give our cells a barrage of sugar. Over time they react to the high levels of sugar by blocking it. This can lead to diabetes. You won't get fat or sick from eating fruits and vegetables.

## *Fruit Juices*

Fruit juice sounds so healthy doesn't it? People who drink juices are getting their fruit right? Wrong, even the so called "healthy juices" that do not add sugars are not the same as eating a piece of fruit.[57] After stripping away the fiber, what's left is a sugar water of sorts that passes directly into our bodies. Again the original fruit that was packaged by God is best. Sorry but we are not there yet. It is really all about packaging. Nature evolved our foods along with us. How many times have you heard "watch your fruits they have a lot of sugar!" It is not as simple as "fruit has sugar". It is the package that counts. The sugar is mixed with proteins and fiber and minerals. As a package our bodies use it wisely. Broken down, it has a negative effect on us. Many juices, especially, concentrates have added sugars as well.[58]

Fruit juice is an unhealthy food. "Even though juice is 'from' fruit, it is not fruit," Lesley Edwards, coordinator of Mission Hospital's Child Weight Management Program said. "All juice has been stripped of the healthy fiber that makes fruit so good for you.[59] We are so marketed that juice is a good thing and healthy but it's not. Juice makes a lot of money for the juice industry but is not healthy and may even be unhealthy for us. Fresh fruit has fiber that helps deliver the natural sugars in fruit into the body slower. Observe that early man had plenty of fruit to choose from but no juice. We evolved on fruit not juice.

If you are one of many people that think they are living healthy and eating healthy but drink juice instead of soda, don't be fooled. Eat your fruits fresh and whole, not processed and bottled. Your body will thank you for it.

## *Grains*

Grains are also essential to life. They provide us with fiber and minerals to help support life. Natural grains are complex and healthy but processed grains (such as white bread) have lost their complexity and benefits. Cereals and breads are carbohydrates and as such are converted to sugars by our bodies. The highly processed ones convert to sugar faster than the whole grain equivalents. This sudden increase in sugar can lead to diabetes and other health hazards.

Americans love their morning cereals. Some of us eat a lot of sugary cereals. Others eat multigrain cereals and believe they are eating a healthy breakfast. Other than raw oatmeal, all cereals convert to a lot of sugar in our bodies. Typical cereals can have 5g to 15g of sugar in a serving. Even the healthy sounding ones can be high in sugars. The oatmeal we eat, a non instant multigrain cereal, has only 1g of sugar in it.

"We know that whole grains are better than refined grains because of fiber, vitamins, and minerals," says researcher Joanne Slavin of the University of Minnesota. Now she and others are beginning to ask whether other things in whole grains-antioxidants, lignans, phenolic acids, phytoestrogens, and other phytochemicals may help reduce the risk of heart disease, cancer, and diabetes. "Like fruits and vegetables, it's the package of nutrients that may be important," says Stavin.[60]

When it comes to health, the food package counts! Make sure your food was packaged by nature not man.

## The Cooking Myth—"I don't have time!"

We seem to have developed a society of not cooking and eating on the run. Cooking has so many aspects to it, beside the preparation of a meal. It produces wonderful smells in the house that kids will remember all their lives. It produces fresher better tastes to be enjoyed by all (It causes a family bonding around the kitchen). So why don't we cook anymore? Mostly because we think it takes too long and we just don't have time. We also don't meet as a family unit as much as we used to. Rachael Ray of the Food Network[61] is famous for her 30 minute meals. She has a large variety of meals that can be cooked start to finish in 30 minutes. These are all healthy meals (with the elimination or reduction of meats). Can you really get the kids in the car, drive to your favorite fast food store, order the food, eat it and drive home in less than 30 minutes?

At home, we are constantly amazed at the meals we cook together. The time is fast, the meals are nutritious and the cost is lower than eating out and provides a bonding opportunity. We enjoy adding up the cost and comparing it to our eating out experiences. We have found our food to be better tasting, the correct portion size and much less expensive. Why are you not cooking more? We are not saying to never eat out. It's actually more fun as a special experience not an everyday experience.

If you are saying "I cook often! I just buy packaged foods that I can microwave or quickly prepare". Read the labels! It is not food; it is chemicals that taste like food. You don't save any time and you will spend much more this way. I was at the house of someone who was preparing mashed potatoes from a box. Come on, the word potato was half way down the ingredient list. How hard is it to boil water with a few potatoes in it and mash them? Is peeling potatoes really that difficult? Buy real potatoes, grown by nature, and cook them with the skin on and mash them with skins for extra fiber.

**Ask yourself one question: "If you don't have time to cook and be healthy now, where will the time come from to take care of you when you get sick later?"**

Is getting sick really worth this type of cooking and eating? Cooking healthy is cheaper, faster and more fun (great tastes and smells). Have you ever walked into someone's house that is a cook? Maybe you smell garlic or fresh basil. Believe us, it really doesn't get better than this. Eating is not only about stuffing our mouths. It is about smelling our foods, seeing them, tasting them and knowing they are healthy for us.

## Other Things We Ingest-The Smoking Syndrome

Although smoking is not technically eating, we do ingest the smoke and nicotine. How many times have you been told about someone that was 90 years old and smoked 2 packs a day for his entire life and never got cancer? It is an intriguing argument until you look at statistics. Most people that smoke throughout their life suffer or die from cancer or other smokers diseases. A very small percentage is unaffected. This is like playing

Russian roulette. You could play a while and not be killed but most that try this game die of a gunshot wound. Remember food is also smell. Does smoke from cigarettes smell good?

Epigenetics is similar. You can eat badly, have stress and live in a poor environment and not be ill. Most people in this situation do get ill. This is clear when we understand just how many Americans are sick and need drugs and hospital stays. Try counting the number of hospitals in a 10 mile radius of your house. Then count the number of cars in the parking lots.

**Food Related Chronic Illnesses**

The following diseases and chronic conditions have been related to food:[62]
- Migraine
- Headache
- Acid Reflux, GERD, or Gastro esophageal reflux disease
- Gastro-Intestinal
- Irritable Bowel Syndrome or IBS
- Constipation or Diarrhea
- Crohn's disease
- Colitis
- Interstitial Cystitis or IC
- Asthma
- Eczema
- Fibromyalgia or FMS and Chronic Fatigue Syndrome or CFS
- Arthritis
- Autism
- ADD or ADHD

As more studies are done, we find food does affect our genes through epigenetics. We know that diabetes is food related as well as many cancers and heart diseases.

## *Science*

Epigenetics and Nutrigenomics will pave the way for a new science that will be able to better understand chemicals in certain foods and how they interact with our cell biology.

The following discussion is from Nutritional Concepts, Inc. [63]

We have always had a unique perspective on chronic disease: in particular, dementia, diabetes, cardiovascular disease, hypertension, and autoimmune disorders: rather than focusing on specific diseases, we feel that in most instances, there is a common thread that ties them together. As nutritionists, we believe that common thread is the fuel we put into our bodies.

Whereas in the traditional diagnostic model, the patient that presents signs and symptoms of inflammation of the joints, depression, headaches, irritable bowel syndrome, high blood sugar, hypertension, cognitive impairment, and skin problems might be seen by a diabetologist, rheumatologist, neurologist, gastroenterologist and dermatologist for each complaint. In our model, we evaluate the interaction of the environment of the patient with their gene potential to discover if there is a unifying theme as the triggering factor. This model now has a scientific term: Epigenetics. It is considered the new frontier in the fight against chronic disease.

In a sense, by preventing physiological imbalance by use of web-like rather than linear cause-effect analysis, we have found that it is possible to address many of the presenting complaints simultaneously with diet and nutrients. While not as fancy as epigenetics, we call it the "whole health approach."

For example, exploring how nutritional environment influences the insulin signaling process takes on a whole new meaning. Recently, it has been brought to the medical community's attention that insulin is more than a hormone that regulates insulin through its influence on genetic expression, but also insulin signaling itself is regulated through a complex network of kinase enzymes that translate information from outside the cell to its interior including the genome and mitochondria. The fact that kinases are intimately affected by dietary influences as well as exercise and stress hormones opens up a brand new science-based approach towards prevention.

The process, simply put, works like this:

1) Food Molecule encounters Kinase Hubs (enzymes)
2A) If food molecule is cytotoxic (items that are toxic to cells), multiple stressor signals are sent to our book of genes
2B) If food molecule is friendly, multiple calming signals are sent to our book of genes

3A) With constant exposure to stressor signals, polymorphisms (defects) that exist in our genes will be expressed negatively.

3B) With constant exposure to calming signals, polymorphisms (defects) that exist in our genes will be latent.

We are just now learning how outside factors like food, stress and environment are affecting the firing of our genes. Are your genes working for you or against you?

**How Taste Works**

It turns out sweet and sour come from our taste buds (on our tongues) but all other tastes come from our noses. We actually smell our tastes not taste them. While food is being chewed in our mouths, odors escape up an opening into our nasal passages and the brain interprets this as tastes. This is why when we have a cold we cannot taste our food.

Remember our tastes come from our sense of smell. This is why a wine connoisseur gargles wine to get the odors into their nasal passages so they can detect minute tastes in the wine.

## *Conclusion*

We like to think about what we are eating and ask the question: "Did early man eat this way?" If the answer is Yes, it is fine to eat but if the answer is no; avoid it with a passion. Below is a table of manmade items vs. God/nature made items showing their calories, Glycemic Index (See Chapter 8: Glycemic Index).

| Item | Manmade | God Made | Calories | Glycemic Index |
|---|---|---|---|---|
| Apple (1 cup) |  | X | 65 | 40 |
| Broccoli (1 cup) |  | X | 31 | 0 |
| Cereal (All Bran—1 cup) | X |  | 160 | 40 |
| Ice Cream (Edy's Grand Vanilla—1 serving) | X |  | 140 | 61 |
| Carrot (1 Cup) |  | X | 52 | 49 |

| | | | | |
|---|---|---|---|---|
| Chocolate Chip Cookie (1 cookie) | X | | 140 | ? |
| Cereal (Corn Flakes—1 cup) | X | | 100 | 92 |
| Bok Choy (70 grams) | | X | 9 | 0 |
| Coca Cola (1 can) | X | | 140 | 77 |
| Doritos Chips (1 Oz) | X | | 150 | 72 |

**Figure 8: Table of Manmade vs. God Made Items**

Recently we saw an announcement from a large donut maker that they were introducing a cheeseburger on two donuts. Do we really need this? Does anyone believe this is healthy for us? Other places also offer burgers on a pizza or burgers on donuts as well. [64]

Thinking about early man as a model is a great way to stay focused on your health. We know you sometimes want some ice cream or a piece of cake or pretzels, etc. These items should be the exception not the norm. Our bodies are wonderfully programmed to keep us healthy and fit. If you mostly eat organic healthy nature made foods, you will be healthy.

**The Singing Scientist**

YouTube has a song by scientist Dr. Matt Barnett from AgResearch in Auckland, New Zealand on epigenetics and food[65]. You can play it and see the words:

*Some folks are short, some folks are tall.*
*Some folks don't have much hair at all.*
*Two things that make this happen, it seems.*
*Well, the first there's food, then there's genes!*
*Now genes are kind of like a plan,*
*to decide whether you're a woman or a man.*
*Then they interact with food, you see,*
*to decide what kind of woman or man you'll be.*
*You might think this is all complicated*
*(how confusing it is can't be over stated).*
*But here's another thing you can take home.*

*There's a piece of the puzzle called your epigenome.*
*Eating for your epigenes, you'll be the best you've ever been.*
*You may not think that it's true, but you can eat your way to a healthy gut!!*
*Epigenetics changes DNA, makes your genes behave in a different way.*
*Like lots of switches, and what's really neat, they turn genes ON or OFF.*
*Depends on what you eat! There's DNA Methylation, Histone acetylation*
*Phosphorylation, even Ubiquination.*
*All are changes that you'll likely find Happen right now in your intestine!*
*We're all unique, Our switches aren't the same.*
*If they're not set right, you could be sick or lame.*
*But the good news is, if our research comes through.*
*We can find the right foods to make a better you!*
*Eating for your epigenes, You'll be the best you've ever been.*
*You may not think that it's true, but you can eat your way to a healthy gut!!*
*Eating for your epigenes, Will make you strong, and lean, and mean.*
*You may not think that it's true, but you can eat your way to a healthy gut!!*

What a great song on exactly what we have been saying! Dr. Barnett is looking at epigenetics and its effect on diseases like Crones Disease and how something like food may influence it through our epigenetics. Early results from his research show that there is a relationship between food and health through epigenetics. It's your choice to be healthy or not. Choose wisely and watch what you wish for!

Food is the most critical thing you can change in your life for a more healthy life. "The food you eat is among the most significant factors affecting your genes and pushing them toward cancer by causing mutation or disruption in their function. That is, what you eat can either prevent cancer and other chronic illnesses or help cause them."[66] Investigate which foods are good for you and which will help with different issues and diseases. Dr. Dharma Singh Khalsa, MD, in his book Food as Medicine, spends a lot of time on this subject. Most of his book is divided into first different foods and what they do and then chronic illnesses and what you should be eating to help your condition[67]. Another similar book is Eat and Heal by the editors of FC&A Medical Publishing.[68]

Remember, it is not doctors, hospitals, acupuncturists, homeopaths, chiropractors, energy workers, or other health professionals who are responsible for our well being. Health begins with us.[69]

If you really want to know what to eat to remain healthy, look at what diets doctors and hospitals place their patients on after they are sick. If you have heart problems or cancer they tell you to cut back on meat, fats, and sugars. Why do so few doctors suggest this before you are sick?

*"Where your pleasure is, there is your treasure; where your treasure, there your heart; where your heart, there your happiness."*—*Saint Augustine* [70]

# Chapter 5:

# Environment—You are what you Think

## Introduction

Merriam Webster defines ***Think*** as:
1. to form or have in the mind
2. to have as an intention
3. to have as an opinion
4. to reflect on : ponder
5. to call to mind : remember
6. to devise by thinking
7. to have as an expectation : anticipate
8. to center one's thoughts on
9. to subject to the processes of logical thought

The book "The Secret"[71] popularized the idea that you can change your world by thinking the changes you want. For most people this is hard to believe. Most people do believe that there is a power in positive thinking. We strongly believe in this. Our free will gives a choice after every event in our lives. Sometimes these events are catastrophic like the loss of a loved one. How can anyone feel positive after that? Psychiatrists tell us we have to go through a healthy grief period. This helps us deal with the pain of our loss. After that we can choose to be locked into our grief and continue to feel bad, seeing everything as negative or we can look for our blessings. Bad things happen to all of us in life but at the bottom of every valley is a new mountain that can take us back up in life. This allows us to feel better about ourselves again. We can choose to be negative or positive. It is our choice.

People that are negative tend to have negative things happen to them in life and positive people tend to get positive rewards. Once again it's our choice to be positive or negative. Our bodies will react to both! Do you want your body reacting to negative thought or positive thought?

## *Basics*

### Negative vs. Positive Thinking

Choosing to be negative does affect our attitude and our health. As The Secret says, we get what we wish for or believe in. An old saying says: "Be careful of what you wish for; it may just happen!" Negative thinking beyond the initial grievance causes some epigenetic change that can affect our health in negative ways. It is like our brains are trying to give us what we want . . . more grief. Dr John Hagelin, Quantum Physicist and Public Policy Expert, says, "Our body is really the product of our thoughts. We're beginning to understand in medical science the degree to which the nature of thoughts and emotions actually determines the physical substance and structure and function of our bodies.[72] Stress can cause us to think negative and that kind of thinking can and will cause us to be sick. Our mind is such a powerful tool that in the wrong hands, it actually will hurt us. Dr. Ben Johnson, Physician, Author, and Leader in Energy Healing, says: "we've got a thousand different diagnoses and diseases out there. They're just the weak link. They're all the result of one thing: stress. If you put enough stress on the chain and you put enough stress on the system, then one of the links breaks".[73]

There are many stories about people that have been down and out and have turned their lives around.

Maybe this affect on our bodies is not a bad thing. After all, if we touch something hot we feel pain and move our hand quickly. This is to protect our hand from a more serious burn. Our body is communicating with us about something we are doing wrong. Maybe our bodies getting sick from stress or bad food are the body's way of saying STOP IT! Can we turn around the sickness?

Dr Michael Bernard Beckwith, spiritualist, says: "When a person has manifested a disease in the body temple or some kind of discomfort in their life, can it be turned around through the power of 'right thinking?' And the answer is absolutely, yes."[74]

We have to train ourselves to listen better to what our bodies are telling us. Do you listen to your body? Is it screaming out at you for change? If so, why are you ignoring it? What would happen if you ignored your body's message about your hand on a hot surface? That kind of pain is hard to ignore but so is pain from serious illness. Start each day by asking yourself "what is my body trying to tell me?" Listen to it; make it your life's work to change things so your body is happy. Listening to your body's messages will make you the happiest in life. You will be healthy and positive. Money isn't what it's about. Health and happiness is. Get seriously ill and see how important money is to you, other than paying for medical treatments that may not help you. After all, health and happiness are conditions of being in sync with ourselves and with nature. Ignoring this state of being will cause stress and illness.

Think positive, it does matter. Our brains want to deliver to us what we want. If you think bad and have negative thoughts; you will get your wish fulfilled. We all know this but in our modern world it seems to be old and cliché. How well do these thoughts fit into your life?[75]

- Making your own choices and living up to them
- Being responsible
- Being inspired
- Have few regrets
- Do what you love
- Do the right thing
- Cultivate positive energies
- Stop complaining
- Accept responsibility
- Be around people you trust
- Never say never

Albert Einstein wrote:

> *"I have never looked upon ease and happiness as ends in themselves . . . The ideals that have lighted my way and time after time have given me new courage to face life cheerfully, have been Kindness, Beauty and Truth."*

When we commit ourselves to live by truth, goodness and beauty—we transform our lives, because what is true, good and beautiful represents universal values worth living for.[76]

How many of us ever even think about Kindness, Beauty and Truth? They seem cliché in our modern world. We do talk about beauty but it is typically Hollywood beauty not deep down soul beauty. Our society has moved away from kindness and we treat people just the opposite. It is all about us. Truth has no meaning in our society. We lie to advance ourselves. What has happened to us? We have lost confidence in big business. Stories like the Enron story are commonplace. We have also lost confidence in our government. It no longer protects us but is a means for politicians to make a lot of money. This is happening more and more because we allow it. We still vote for politicians that don't have our best interest at hand and we buy from big businesses that only have their compensations in mind.

**Figure 9: Is the Glass Half Empty or Full?**[77]

## *Advanced*

**Forgiving the Unforgivable**

It is so hard to forgive someone that has done an atrocity to you. Think of your ex-spouse in a divorce, an old lover, someone that hurt your child, someone that killed a friend or loved one. How can we begin to forgive such evil? The answer is that forgiveness is not about the other person! It is all about us. To move on in your life you must be able to forgive. If we don't forgive our minds are stuck in a hatred that will bring us down. Forgiving is not about letting the other person off the hook. It is about ending our hatred and negative thinking and moving on with our lives. We all have things worth living for. We all get blessing in our lives no matter how bad things are for us. You just have to look for the blessings. Holding onto this kind of negative energy causes epigenetic changes that can cause unhappiness and illness.

Jo Anne's first husband, Gail, was diagnosed with ALS (Lou Gehrig's disease). He knew he would die but refused to be consumed by that thought. He made, along with his church, several videos about how he felt. He spoke about love and truthfulness.[78] How much love do you have? Jo Anne sees rainbows all the time now. She believes Gail is sending her a blessing that all is OK. Are your eyes wide open and looking for your blessings or are they shut tight? Going through life with our eyes shut tight can only cause us problems and unseen dangers.

Can the act of another easily destroy your love? Beverly Flanigan speaks about overcoming the bitter legacy of intimate wounds.[79] Beverly speaks of the phases of forgiving:
1. Naming the Injury
2. Claiming the Injury
3. Blaming the Injurer
4. Balancing the Scales
5. Choosing to Forgive
6. The Emergence of a New Self

If no one forgave anyone; our society would fail. We would be a race of people hating everyone else. Not forgiving, pure and simple, is not a healthy way of life. Forgive, forget and get on with your life. Again it's your choice. Free will is not so simple or easy is it? Each of us makes billions maybe trillions of decisions in our life that affect us. Some of these decisions will adversely affect our health. If you have made some bad decisions in your life; you can always change them. This change will make you get healthier and happier.

Time is the present! The past has happened and can't be changed and the future doesn't exist (at least the way we think of it). The future is really a vast set of possible courses one can take. The one we actually take becomes the new present time. Our futures are not fixed, so don't think of them that way. No matter how bad your life feels, each day gives you an infinite number of possibilities for a better future. You just have to choose the correct path. We really only have the present. Each day gives you another chance to correct your way of life. Doing so in a positive way will affect the future decisions you are faced with. Remember the old Scottish folk song "you'll take the high road and I'll take the low road?" Life is like that. We can take the upper road on a positive note and be happy or decide to take the low road and be negative and unhappy. Which road are you on? Which one do you want to be on? If the answer to these two questions is different, change your life.

**The Cost of Negative Thinking**

It costs too much, in happiness and health and money, to be negative. You gain nothing in being negative except maybe a false boost to your ego. Our egos want to control us. It is the part of our brain that causes us to feel apart from others. It causes us to be an individual. When our egos get too over inflated, they begin to make us feel better than others. We are smarter, richer, better looking, etc. Learn to keep your ego in check. Ask yourself if your feelings are the way you think or just your ego. You can ignore your ego. Happiness and health cannot be bought, they don't come only to the rich or smart people and they don't come to super models only. In fact studies show that rich people, very smart people and super models are not always the happiest or healthiest people around.

Sigmund Freud believed our brains were comprised of three parts 1) the superego, 2) the ego and 3) the Id. The Id is part of the unconscious mind that deals with instincts, desires, aggression and sexuality (primitive drives). The superego represents our conscious minds. It counteracts the Id with a primitive sense of morality. Finally; the ego is part of both the conscious and unconscious minds. It stands between the Id and the Superego. It includes our conscious sense of self and the world around us. It also consists of our unconscious control of inhibitions, character and personality.

We all have egos but we also have free will and can choose to ignore our egos or at least control them. In doing so we look at all life as being worthwhile, not just our own. We see ourselves as part of a wonderful body of life not just an entity competing for position. We have spoken about how our epigenetics interacts with everything to adapt us to developing situations. Ego is no different. Epigenetics must deal with ego and its ramifications. Some of this is good for us and others can be harmful to us. If we have a very strong ego, we will cause different epigenetics to enable and disable our DNA. These differences can cause major differences in our health and well being. Remember we said our bodies are a community of individual living cells. They change all the time. So we really aren't an individual but a society grouped together for survival.

**Can Epigenetics build better Humans?**

This is a dangerous area of exploration. Can we ever return to the lower stress levels of early man or eat like we did back then are goals that will help us. It took our bodies millions and DNA billions of years to evolve. We evolved in an environment that we had to deal with. There was stress but it typically went away when the event causing stress went away. Today we live in a state of constant stress. These modern changes happened in a very recent biological timeframe. Our bodies and DNA had no time to react but our epigenetics is a system that tries to react in real time. To blindly try to re-program man to be better is very dangerous. Without full knowledge of DNA and epigenetic codes, we would be shooting in the dark.

"Neutralize the elements of human nature in favor of pure rationality, and the result would be badly constructed, protein-based computers."[80] The apparent imperfections in human nature, like adolescent violence, may spring from the same epigenetic rule that guides explorers and mountain-climbers, he suggests. Any attempt to change the apparent imperfections in human nature "would lead to the domestication of the human species—we would turn ourselves into lapdogs." [81]

We need to be very careful when the programming we are changing is what makes us human.

## *Science*

**What Science says about it**

Physicists know in Quantum Mechanics there is a now famous experiment known as the Double-Slit Experiment or originally known as Young's Experiment. This experiment proves the Particle-Wave Duality (or just Duality) property of Quantum Mechanics. This property states that if we observe things in nature, the act of observation changes the things we are observing. WOW, think about that. We have the power in our minds to change what we see. This experiment was to measure if an electron acted like a wave or a particle. It turns out if we do not try to measure it (observe it) it acts like a wave but as soon as we try to measure it, it acts like a particle. This is hard to understand and takes a bit of faith.

So modern physics has an example of what The Secret spoke about. We do have powers to change what happens to us and how we feel about our lives. It doesn't mean you simply think about winning the lottery and it happens. We can choose to be positive and have that energy bring about positive things in our lives. The duality experiment brought Physics and Philosophy together on this point.

Epigenetics is showing us that how we think can change which DNA genes are enabled. This in turn can affect our health and well being. We are just beginning to study these affects and what they can do to us. The science involved here will expand drastically over the next 10 years.

## Conclusion

We saw a great quote outside a church that said: ***"Worry is the darkroom in which negatives are made."*** How much of our negatives are brought on by us. We have known couples that have lost children. This is a pain that must be devastating. Some never got over it and became ill themselves while others seem to be able to grieve and then get on with their lives. We are not saying that they stop thinking about their loss or how much they miss their lost child but it doesn't control them or consume them. If you are religious and believe in an after world, then how can dying and being with God be a bad thing? Our feelings about a lost one are selfish. We just miss the dead person so much.

Don't die wishing you had said something to someone and didn't. Never go to bed angry and don't get caught dying while you hate someone. Life is too short to let our human egos interfere with our happiness. Swallow your pride and take the step that brings you back in touch with your loved ones.

We can tell you for sure that on your death bed you won't be:
- Worrying about a sale
- Thinking about your job
- Wondering about things lost
- Wanting to know how much interest you made this month
- Wondering who won the Super Bowl

Thinking positive and banishing anger and hatred from your life will make you happier and it's healthier!

The Dalai Lama said: *"It is under the greatest adversity that there exists the greatest potential for doing good; both for one self and others."* Maybe it's time for Americans to take control and change their lives for the better. Think about what happened in New York City after 911. People came together to help others. They consoled those that lost loved ones. An entire city with a reputation for being cold; cried together, helped each other, and yes, laughed together. The cycle of life went on. For just a brief moment the entire country was tied together with one another again. Do you see that even in such a horrible disaster, blessings were abundant?

This chapter and the next are really about how to be truly happy in life. Happiness is not based on material things alone. Robert Louis Stevenson said "There is no duty we so much underrate as the duty of being happy." You always need to ask yourself "Are you happy?" If the answer is no, find out why, change your life so you are happy. Your cells are depending on it. Gretchen Rubin said[82] "Time is passing and I am not focusing enough on the things that really matter." In that moment, she decided to dedicate a year to her happiness project. Her book, The Happiness Project, shows how to build your own Happiness Project.

Thinking is a great gift given to man. It also is a great curse if used wrong. Thinking over and over about something you can't change, can raise your blood pressure and harm you. Meditation is an exercise that can help break this cycle of thinking and bring peace to our inner soul. Use your gift of thinking wisely!

# Chapter 6:

# Environment—You are what you Believe

## *Introduction*

Merriam Webster Dictionary defines belief as:
1. a state or habit of mind in which trust or confidence is placed in some person or thing
2. something believed; *especially* : a tenet or body of tenets held by a group
3. conviction of the truth of some statement or the reality of some being or phenomenon especially when based on examination of evidence

Belief in something is a basic human trait. George often uses an example of true belief as driving up and over one of San Francisco's numerous hills. As your car goes over the top all you can see is sky. You continue for one reason only! You believe without a doubt that the road continues on the other side of the hill. You can't see it until the car is over the crest and starting down the hill. If the road was not there, it would have been too late.

We all need to believe in God, family, friends, our country, etc. It is our nature to want to believe. If you are not religious; then belief in God is out. If your family is estranged then belief in family is gone. If you don't trust anyone then belief in friends, country, etc. is gone. Reconnecting to people and God brings back a basic need in all humans.

## Basics

### Religion

We see positive thinking and believing as related. It doesn't matter what form of religion you practice. If you believe in God, you must view Him as a positive and loving creature. He wants you to be positive and loving as well. He has created our bodies, our DNA and our epigenetics to give you that wish, if you want it (free will). Do you want it? How badly?

Belief is an intrinsic part of human life. We believe in family, friends, relationships, our government and our way of life. When things happen that dispel those beliefs, we have stress. We can choose to learn from the experience and move on with a happy life or dwell on it and live unhappily. The latter will probably bring with it an unhealthy life.

We all know people that are very involved in their way of thinking about politics. When the country goes against their politics it usually causes them great stress. You often hear things like "Oh no, what are we going to do now?"

What we eat, think and believe has always been important to humans. There now appears to be an epigenetic link to each of these. *Dawson Church, PhD, in his book The Genie in your Genes*[83]*;* says epigenetics can affect how we live and to what age. Factors that affect this are:
- Having a Body Mass Index below 25
- Eating a diet rich in fruits and vegetables
- Daily aerobic exercise
- Our beliefs
- Feelings
- Prayers, and
- Attitudes

Dr. Church says over 1,400 chemical reactions and over thirty hormones and neurotransmitters shift in response to stressful stimuli. This says that with every feeling and thought, in every instant, you are performing epigenetic engineering on your own cells, [84]

Are you performing good or bad epigenetic engineering on your cells? If your engineering is off you will suffer with stress, weight and health issues. If on the other hand your engineering is good, you will be happy, of an ideal weight and healthy. Both mind and body benefits from good engineering. You already have all the tools you need to self heal. God gave them to you along with external items we need such as minerals, fruits and vegetables.

Each of us can eliminate or block those inputs that cause us harm. It is not always easy to do so but it is within our power. You can move to a cleaner and safer city to live. You can choose to eat more fruits and vegetables. You can change your job and have less stress on yourself. You can love your work and work your love. You can reduce the amount of processed foods you eat. You can believe in God and people again. Epigenetics is the engineering of self choice (free will) on our bodies and minds. When really bad things happen it was designed to preserve at least some of the life. Imagine the meteor that wiped out the dinosaurs, it killed a lot of life on Earth but not all of it. Epigenetic changes from the environment caused DNA to modify what life was. Some of the modifications didn't work out and died off like the dinosaurs while others were given a change to flourish, like the mammals.

For all of mammals' existence and most of human existence, these changes were reactions to available food, weather, natural disasters, etc. But in very recent times, human epigenetics is reacting to what we have done to the environment, to processed foods, changes in our diets, to increased stress levels and to how we believe. These reactions have been mostly bad ones. We are getting sicker and unhappy as we evolve. Will these changes mark the end of humans like the dinosaurs? Who will be next in line to dominate the Earth?

Figure 10: Religions[85]

## *Advanced*

### Our Spirits and Life

Each of us has a soul or a spirit. We believe that the spirits of all living creatures are linked together. Life here on Earth is like a virtual reality ride. We are in a dreamlike state where evil can exist as well as hardships and pain. Each of us has an infinite number of paths through this life. Some will bring us much joy and others much pain. We choose the paths we take with every decision we make along the way. Some are hard to control like making a decision to go to the store and having an auto accident on the way. Had we not decided to go to the store we would not have had the accident. None of us has a crystal ball to see into the future and determine which path is best for us. We can make as many intelligent decisions as possible in our lives. Life is a series of up and downs, good and bad, evil and love. We must know hate to appreciate love. The Chinese called these opposing forces, Ying and Yang. The Ying and the Yang explains these differences in our lives.

Our spirits are like being on the Hallo Deck of Star Trek. We are visiting this place we call Earth. This is an illusion for the spirit world. The spirit lives outside our 5 senses. The Hallo Deck brings the illusion alive and reality via our 5 senses. Have you ever woken from a dream that seemed so real? It isn't until after awakening that you realize it was only a dream. Death may be like this. When we die we pass back to the spirit world and realize this has all been an illusion. Death may be no different than awakening from a dream.

The idea of life being like a video simulation game was talked about by Morgan Freeman in his new Science Channel Series, "Through the Wormhole".[86] Morgan Freeman introduces Will Wright, the creator of the blockbuster game "the Sims". He explains our universe and life may be nothing more than a simulation created by God or an advanced mind. If you look at a video game only the current frame is real. The other frames have not been created until you navigate there. Life is the same says Quantum Mechanics Physicists. Quantum Mechanics shows that things are as they are only when we observe them. Sounds very much like a simulation game. If we are inside someone's game, who created us and where is this creator? It may just be that our bodies are created in this simulation to show our souls what life without infinite love is like.

"Do you think that anyone can damage your soul? Then why are you so embarrassed? I laugh at those who think they can damage me. They do not know who I am, they do not know what I think, they cannot even touch the things which are really my own and with which I live."—Epictetus [87]

If you believe that you are mostly your soul, your spirit and not your body, then nothing can harm you. Things in this world can only touch or harm your body. The spirit is of another world that cannot be touched. No matter how bad things are for you here in this world, your spirit lives in love in another. Count your blessings.

**Our Senses**

We are physically "plugged into" this world through our 5 senses. The 5 senses are:
1. Sight
2. Hearing
3. Smell
4. Taste, and
5. Touch

**Figure 11: The 5 Senses**[88]

Try to imagine not having any of these senses. We don't believe you would be alive. You would know nothing of this world. You could not interact with it in any way. This virtual reality life we live is controlled by our senses. We can lose one or two and have the others become stronger but we cannot live without any of them. A person without senses is not really a person. You would feel no pain, know no one, eat nothing or not know you were eating, see nothing. Nothing as we know it would be available to you. Dean Koontz in his Frankenstein series: Lost Souls says: "If humanity no longer exists on Earth to see, hear, smell, taste and touch the abundance of nature, does Earth itself continue to exist in its absence? His answer is no. The mind perceives matter and invests it with meaning. Without the mind to observe it, matter has no meaning; what cannot be perceived by any of the five senses—does not exist."[89]

The idea that we were somehow randomly made by random changes to our DNA is rapidly dissolving. We are a much more programmed species that interacts and changes with everything we encounter. We are the product of an intelligent being just like programs on computers are the product of intelligent programmers. A computer's memory cannot just randomly change such that software like Microsoft Windows is suddenly created out of nothing. It took hundreds of human programmers to develop it through their combined intelligence.

## Man Made Sensory Control

Imagine we humans invent a machine that can perfectly stimulate our five senses. It would control:
1. What we see
2. What we hear
3. What we smell
4. What we taste
5. What we feel

If we were placed in the control of this machine, we would be in another reality. The 2010 movie, Inception, was about this type of dream state. Our minds would believe it was real. We would not be able to escape it until the machine was disconnected or turned off. The movie **The Matrix** was another example of this. Neo wakes up in his pod and realizes everything he knew and believed in was manufactured by the machines. His body is really a battery to fuel the machines. They give the humans the made up reality to control them. Our life here may be very similar to this. God may have given us our five senses and a world to control them to teach us what it would be like to be in a world of non infinite love. He may be allowing us to experience a world of hatred and evilness so we can appreciate love.

## God is all Loving

We believe in a God that infinitely loves. The physics education George had gives him a good understanding of infinite. If God infinitely loves, then he cannot hate, be bad or be prejudiced. We also believe that this type of God exists in a place that has no evil, no hardship, and no hatred. It can only be a loving place. God only knows why we are here. Maybe it's to understand what a non-loving world would be like. He did not put us here and abandon us. He gave us the food, the air, the water and the sun we need to exist healthy. Are we really using the tools God gave us?

## Physics, Good & Evil

The science of physics shows us that things we truly believe in don't really exist. A great example is the hot vs. cold. Cold does not exist. What we call cold is only the absence of heat. This is why, when we remove 100% of the heat from a container, we have a temperature of Absolute Zero. It cannot get any colder. You cannot remove more heat than there is in the container. It may very well be that evil doesn't exist either but is the absence of love. Light and darkness are similar concepts. There is no such thing as darkness only the absence of light. If you remove all the light from a container it is totally dark. It cannot get any darker. We have flash lights to make an area lighter but there is no such thing as a dark light to make a lit room darker.

## Religion—Food for Thought

If you are religious and a Christian, you know Jesus said we all can do the same miracles He did; if we believe. He also said we could move mountains if we had the faith of a mustard seed. We have incredible brains there are capable of so much more than we use them for. We know our universe is made up of energy. The latest Physics theory (String Theory) states that at the bottom of each sub-atomic particle is a series of vibrating strings. Pure energy that determines which particle is created by their vibration frequencies. We also know our minds give off energy as part of our thinking process. We also know our cells react to our thoughts via epigenetics. Maybe all of these mechanisms are the foundation for what Jesus said.

Religion gives us something to believe in. Belief is a basic need we have. We were created and evolved as creatures that ate properly, had occasional stress but then let it go, believed in our world, our God and our environment. In the last 100 years or so we have drastically changed what we eat, have unending stress and don't believe in much except ourselves and greed. We are seeing the price of such change.

## *Science*

Science seldom meshes with spirituality or religion well. We may be seeing a new era where science begins to explain what spiritualists and the religious always knew on faith. Physicists studying a new set of formulas known as "String Theory" see a powerful, almost infinite force at the basis of all matter. This force vibrates at different frequencies, like strings on a piano. Each frequency builds a different type of matter. Could this powerful base to everything we know be God? Science and religion will never agree with each other by definition. Religion is concerned about things relating to God. God by definition exists in another dimension. One that is very different from ours. Our Universe is designed to one day end. God is infinite and therefore cannot be of our Universe. Science is concerned with things of this Universe. How do they work? How do they interact? How can we predict what they will do better? The two disciplines are of two very different existences. We believe they can co-exist peacefully.

## *Conclusion*

Belief is basic to humans. They can bring us pleasure and happiness or sickness and unhappiness. The choice is ours. The key idea is we do have a choice.

We have to be observant and watch out for certain things:
- Watch out for the marketing buzz. Understand it's about making someone else rich not about what is necessarily what you believe in
- Watch out for false facts like the food pyramid, government mandates on health and what other people believe like "You need your meat for protein"
- Watch out for false logic like I know someone that lived until they were 90 and smoked/ate badly/were stressed and were healthy. The odds are you won't be that lucky
- Watch out for a life of lost opportunities through negative thinking instead of positive thinking

We can:
- Change our life and be more healthy and positive,
- Use our free will to modify our bad habits,
- Enjoy a healthy diet,
- Get all the protein we need from sources other than meats,
- Ignore the draw of marketing and see it for what it actually is,

- Wean ourselves off most drugs with our doctor's help and a lifestyle change,
- Listen better to what our bodies are telling us, after all, our bodies have been doing this for millions of years.

Believe in this and make it happen.

Who we are is really a summation of the choices we made in our lives. Death ends the possibility of making choices but while you are alive you can still choose. Choose wisely and may the force be with you.

# Chapter 7:

# Vegetarianism

## *Introduction*

Millions of people have embraced vegetarianism in one form or another throughout history. Millions of people in India, China and Japan have been healthy vegetarians for thousands of years. Plato, Plutarch, Isaac Newton, Ben Franklin, Albert Einstein, Leonardo da Vinci, Leo Tolstoy, Albert Schweitzer, George Bernard Shaw, Martin Luther and John Wesley promoted a vegetarian diet. Wikipedia has a list of vegetarians by country. Go to www.wikipedia.org and look up *List* of *Vegetarians.*

Vegetarianism is the practice of following a plant-based diet including fruits, vegetables, cereal grains, nuts, and seeds, with or without dairy products and eggs. A vegetarian does not eat meat, including: red meat, game, poultry, fish, crustacea, shellfish, and products of animal slaughter such as animal-derived gelatin and rennet. There are a number of vegetarian diets. A lacto-vegetarian diet includes dairy products but excludes eggs, an ovo-vegetarian diet includes eggs but not dairy products, and a lacto-ovo vegetarian diet includes both eggs and dairy products. A vegan diet excludes all animal products, such as dairy products, eggs, and honey. Vegetarianism may be adopted for ethical, health, environmental, religious, political, cultural, aesthetic, economic, or other reasons.

Semi-vegetarian diets consist largely of vegetarian foods, but may include fish, poultry, dairy products, and eggs. With these diets, the word "meat" is often defined as only mammalian flesh. A pescetarian diet, for example, includes fish but no meat. The common use confusion between such diets and vegetarianism has led vegetarian groups such as the Vegetarian Society to note that such fish or poultry-based diets are not vegetarian, because fish and birds are animals.[90]

We are promoting much more fruits and vegetables than your typical America consumes. Plant based foods have shown over and over again many health benefits and have considerable disease fighting abilities. Even if you can't go full vegetarian, try increasing the amount of plant based foods and decrease the amounts of animal based foods.

## *Basics*

### Reasons for a Vegetarian Diet

People usually choose a vegetarian diet for 1) health, 2) animal rights, 3) the environment or a combination of the three. In this book we are focused on the first aspect only.

Studies of vegetarian groups have shown a lower incident of heart disease, stroke, diabetes, colon and other cancers, osteoporosis and obesity than in non vegetarian groups.[91]

Wow, is that worth changing your dietary style over? Our epigenetics are so powerful that they can cause drastic changes in our life and health by changing something as simple as diet.

Research has shown that vegetarians are 50 percent less likely to develop heart disease, and they have 40 percent of the cancer rate of meat-eaters. Plus, meat-eaters are nine times more likely to be obese than vegans are.

The consumption of meat, eggs, and dairy products has also been strongly linked to osteoporosis, Alzheimer's, asthma, and male impotence. Scientists have also found that vegetarians have stronger immune systems than their meat-eating friends; this means that they are less susceptible to everyday illnesses like the flu.[7] Vegetarians and vegans live, on average, six to ten years longer than meat-eaters.

A plant-based diet is the best diet for kids, too: Studies have shown that vegetarian kids grow taller and have higher IQs than their non vegetarian classmates, and they are at a reduced risk for heart disease, obesity, diabetes, and other diseases in the long run. Studies have shown that even older people who switch to a vegetarian or vegan diet can prevent and even reverse many chronic ailments.[92]

Imagine how powerful you are, you can change your life by changing the amount of plant based foods in your diet.

The China Study by T. Colin Campbell, PHD and Thomas M. Campbell II is the most widely done study on the influence of diet on human health. This is a very large study that indicates cancer and heart disease are directly linked to eating meat. The authors were able to turn cancer on and back off again in lab rats by either feeding them meat or restricting their diet to no meat. Both sets of rats were subjected to radiation to induce cancer. 100% of the meat eaters died from cancer while 0% of the non meat eaters died or contracted cancer. Rats are used since they have systems similar to humans and are as susceptible to cancer as humans are. They also did human studies on a group of people that were diagnosed as having incurable heart disease. They were not given long to live. Modern medicine wiped their hands of them and said there was nothing more they could do. The authors took the group and over a long period of time improved their health by eating only a vegan diet. They were all alive and doing well at the time the book was published. Their heart disease seemed to have disappeared. In some cases were they had severely blocked arteries, their bodies built new arteries around the blockage.

If you want to read and study this work, we recommend getting it as a book on CD disk and listening to it. The book can be technical and hard to read. The narrator has a very nice voice and makes it enjoyable to listen to. If you are going through cancer or heart disease or know someone that is; we highly recommend you give them a gift of this work. It could save their life. There is no downside. Eating vegetables will NOT kill you.

**Milk, Apple Pie and Motherhood—The American Way**

In our last chapter, we discussed beliefs. What I am about to say will cut to the core of our American way of "wholesome" living, but bear with me on this, as I tell you that two of the above "staples" of life are actually not good for you.

As, you may have already guessed, Motherhood is not in the scope of this chapter, although it is recognized that the type of mothering we each received is an epigenetic factor in our lives. Most of us have been fortunate to have received good or great mothering, so we will allow Motherhood to keep its place as a wholesome American standard!

Apple pie, baked with the usual ingredients, generally has a HUGE amount of sugar, not to mention HUGE amounts of saturated fats in the crust. Certainly, store bought pies are some of the worst. You have no control over the ingredients or amounts. Why do you think they taste so good? Generally, and unfortunately, the answer is, because they are loaded with sugar and fat! An apple pie that you bake at home with care taken regarding the ingredients, can be an altogether different matter. Apples are sweet, by nature, and do not need extra sugar dumped all over them to make them into a pie. Do you dump sugar on your apple before you take a bit out of it? Too much sugar is bad for you and can lead to obesity, diabetes and cancer. Sugar from nature's fruits, like whole apples (not apple juice) is good for you. It is delivered in your body along with the fiber you were also intended to have. Why have we allowed man to tell us that processed "fruit-based" foods they make are better for us and better tasting to us than the food God created?

Kim Barnouin, in her book Skinny Bitch—Ultimate Everyday Cookbook, wrote a recipe for Peach Crisp (we know it's not apple pie but you can substitute apples for the peaches). This recipe has no added sugar.[93] We can eat a limited amount of desserts, as long as we watch the added sugars and use ripe organic fruits that are naturally sweet.

Do you believe that you can never give up sweets? Dr. Barnard has a video on YouTube that discusses addictions. He explains how in 3 weeks our taste buds can be re-programmed to new tastes. Remember when you switched from whole milk to skim? The skim tasted light and watery. After drinking skim milk for a time; drinking whole milk felt heavy and thick.[94] Our taste buds can be re-programmed so healthy foods actually taste good to us and unhealthy foods no longer taste good. Fatty fried foods, sugary sweet foods, and greasy meats were learned by us to taste good. Have you ever seen a baby eat their first fruits or vegetables? They love the taste. It is a natural taste we have enjoyed for a long time.

Ok, you guessed it. This section is really about dumping on milk. MILK, the drink that our mothers always told us to drink because "it's good for us"! Well, sorry to those moms who were "milk promoters", we realize that you were only passing this information on to us in good faith, because this is what you believed to be true.

Dr Bernard's video, also discusses the fact that milk has natural opium in it to make a baby want to feed on it. If nature hadn't done this, we might not have nursed as well and that would lower our chance of survival. Nature never intended us to continue nursing into old age, but we drink milk until we die.

Milk sounds so American . . . .

We believe in motherhood but apple pie and milk can make you sick. Apple pie can provide us with too much sugar which can lead to diseases such as diabetes and cancer as well as obesity. What can be wrong with milk? Think about why nature produces milk in mammals. It is to feed newborns (not adults) and provide them with the protein they need to survive and grow before they can consume solid foods. Human milk was designed for humans. Cow's milk was designed for cows not humans. Cows have two stomachs and crow very large very rapidly. Cow's milk was designed with this in mind. Humans have only one stomach and do not grow as large or as fast. Why would we ingest such a product? No mammal, except humans (mostly American humans), drink milk after early childhood. Milk's proteins are designed to interact with a cow's DNA to trigger epigenetic changes beneficial to a cow. Key words here are COWS DNA. Milk may in fact be far more dangerous than we could imagine. It may be linked to cancers especially breast cancer and other diseases.[95] We are also the only mammal that regularly drinks the milk of another animal.

Scientists are finding that human milk contains a vast amount of different elements that have an effect on our development. These were developed by nature over millions of years as our bodies were developed. Each component in human milk is there for a reason. Scientists found a sugar in human milk that is not digestible by the human body but it coats and protects the stomach. Dr. Mills of the University of California at Davis said: "So for God's sake, please breast feed."[96]

## Are Humans Carnivores?

Carnivores are meat eating animals, herbivores are plant eating animals and omnivores eat mostly plants with some meat. Mammals have been eating a variety of fruits and vegetables for millions of years. Yes some are carnivorous and eat meat. The question is: are we more of an omnivore or a carnivore? Humans are really omnivores. We have historically eaten mostly fruits and vegetables with a small amount of meat here and there. It took us a long time to reach the top of the food chain. Early man most likely got eaten more than they ate other animals. We have a digestive system that is more adapted to plants than animals. We have very small canine teeth and no claws at all.

A very funny book that discusses these concepts in an irreverent but attention grabbing way is "Skinny Bitch".[97] The cover describes this book as "A no-nonsense, tough-love guide for savvy girls who want to stop eating crap and start looking fabulous!" This book is not just for women!

The prominent Swedish scientist Karl von Linne states, "Man's structure, external and internal, compared with that of the other animals, shows that fruit and succulent vegetables constitute his natural food. Table 12: Comparing Meat-Eaters, Herbivores and Humans (*based on a chart by A.D. Andrews, Fit Food for Men, (Chicago: American Hygiene Society, 1970))* below shows a comparison"[98]

When you look at the comparison between herbivores and humans, we compare much more closely to herbivores than meat eating animals. Humans are clearly not designed to digest and ingest meat.

- *Meat-eaters*: have claws
  *Herbivores*: no claws
  *Humans*: no claws

- *Meat-eaters*: have no skin pores and perspire through the tongue
  *Herbivores*: perspire through skin pores
  *Humans*: perspire through skin pores

- *Meat-eaters*: have sharp front teeth for tearing, with no flat molar teeth for grinding
  *Herbivores*: no sharp front teeth, but flat rear molars for grinding
  *Humans*: no sharp front teeth, but flat rear molars for grinding

- *Meat-eaters*: have intestinal tract that is only 3 times their body length so that rapidly decaying meat can pass through quickly
  *Herbivores*: have intestinal tract 10-12 times their body length.
  *Humans*: have intestinal tract 10-12 times their body length.

- *Meat-eaters*: have strong hydrochloric acid in stomach to digest meat
  *Herbivores*: have stomach acid that is 20 times weaker than that of a meat-eater
  *Humans*: have stomach acid that is 20 times weaker than that of a meat-eater

- ***Meat-eaters***: *salivary glands in mouth not needed* to *pre-digest grains and fruits.*
  ***Herbivores***: *well-developed salivary glands which are necessary* to *pre-digest grains and fruits*
  ***Humans***: *well-developed salivary glands, which are necessary* to *pre-digest, grains and fruits*

- ***Meat-eaters***: *have acid saliva with no enzyme ptyalin* to *pre-digest grains*
  ***Herbivores***: *have alkaline saliva with ptyalin* to *pre-digest grains*
  ***Humans***: *have alkaline saliva with ptyalin* to *pre-digest grains*

**Table 12: Comparing Meat-Eaters, Herbivores and Humans**
(*based on a chart by A.D. Andrews, Fit Food for Men, Chicago: American Hygiene Society, 1970*)[99]

There is no more authoritative source on anthropological issues than paleontologist Dr. Richard Leakey, who explains what anyone who has taken an introductory physiology course might have discerned intuitively—that humans are herbivores. Leakey notes that "you can't tear flesh by hand, you can't tear hide by hand . . . . We wouldn't have been able to deal with food source that required those large canines" (although we have teeth that are called "canines," they bear little resemblance to the canines of carnivores).

In fact, our hands are perfect for grabbing and picking fruits and vegetables. Similarly, like the intestines of other herbivores, ours are very long (carnivores have short intestines so they can quickly get rid of all that rotting flesh they eat). We don't have sharp claws to seize and hold down prey. And most of us (hopefully) lack the instinct that would drive us to chase and then kill animals and devour their raw carcasses. Dr. Milton Mills builds on these points and offers dozens more in his essay, "A Comparative Anatomy of Eating."

The point is this: Thousands of years ago when we were hunter-gatherers, we may have eaten a bit of meat in our diets in times of scarcity, but we don't need it now. Says Dr. William C. Roberts, editor of the *American Journal of Cardiology*, "Although we think we are, and we act as if we are, human beings are not natural carnivores. When we kill animals to eat them, they end up killing us, because their flesh, which contains cholesterol and saturated fat, was never intended for human beings, who are natural herbivores."

Sure, most of us are "behavioral omnivores"—that is, we eat meat, so that defines us as omnivorous. But our evolution and physiology are herbivorous, and ample science proves that when we choose to eat meat, that causes problems, from decreased energy and a need for more sleep up to increased risk for obesity, diabetes, heart disease, and cancer.[100]

Human beings have the gastrointestinal tract structure of a "committed" herbivore. Humankind does not show the mixed structural features one expects and finds in anatomical omnivores such as bears and raccoons. Thus, from comparing the gastrointestinal tract of humans to that of carnivores, herbivores and omnivores we must conclude that humankind's GI tract is designed for a purely plant-food diet.[101]

It would seem that we want to be meat eaters so badly but we really were not big meat eaters through history. The table above shows how we are much more like herbivores than like carnivores.

**Human Dietary Change**

Homo sapiens are omnivorous and opportunistic in their diets. They exhibit great regional diversity in their food preferences.[102] In America, our diets have changed radically since World War II. This is a very short timeframe and not enough time for evolution to help us out. These changes have caused wide spread diseases to become common and some even believe crime was affected by diet.[103]

The main findings in a 7-year blood pressure follow-up study of middle-aged employed men are as follows: 1) Higher intakes of vegetables and of fruits were related to less of an increase in SBP (systolic blood pressure)and DBP (diastolic blood pressure) over time, independent of age, weight at each year, and intake of other foods; 2) men with a higher intake of red meat (beef-veal-lamb and pork) had a significantly greater increase in blood pressure; 3) men with a higher poultry intake had a significantly greater annual increase in blood pressure, independent of other factors; and 4) men with a higher fish intake tended to have less of an increase in blood pressure. [104]

Plant biodiversity is essential to human health. Plants provide sources of both nutrients and medicinal agents, form components of robust ecosystems and contribute to sociocultural well-being. Traditional values and scientific conceptions concur on the necessity of dietary diversity, particularly of fruits and vegetables, for health. In the face of economic and environmental changes, increased simplification of the diets of large numbers of people to a limited number of high-energy foods presents unprecedented obstacles to health. Cultural knowledge of the properties of plants erodes at the same time. Conservation of biodiversity and the knowledge of its use, therefore, preserve the adaptive lessons of the past and provide the necessary resources for present and future health. [105]

Plants are so important to our survival, happiness and health. Modern man is on a slippery slope rapidly moving away from plant based foods. Chemical based foods, highly processed foods, cooked foods and DNA altered foods should not be our choice of a healthy diet. Another study shows men that have lost a number of teeth; eat far less fruits than men with no tooth loss.[106] Yet another study, at the University of California in San Francisco, shows a Vegan diet helps fight cancer by changing the genes that are expressed (epigenetics). [107]

Whether you are Vegan (no animal products at all), vegetarian, cutting back on meats, cutting back on red meats or a fully fledged carnivore, vegetables are important to human health. There are many articles recently about cutting back on the amount of meat we eat, especially red meat. What we don't hear enough except maybe from our mothers is that we need to eat more fruits and vegetables. The typical American diet has been getting unhealthier in several ways:

1. increased portion sizes
2. increased amounts of meats
3. decreased amounts of fruits and vegetables (juices don't count)
4. increased amounts of preservatives in our foods
5. the food we eat is getting farther away from fresh cut
6. increased amounts of sweets and carbohydrates that quickly convert to sugars in the body

Many foods are designed by manufactures to be fattening. "Betcha can't eat just one" may be a Frito-Lay marketing slogan for potato chips, but, as any dieter can tell you, most junk food creates addict-like behavior—and not just in adults. [108]

## Healthy Eating

"Eat your fruits and vegetables." You've likely heard this statement since childhood. Research shows why it is good advice:
- Healthy diets rich in fruits and vegetables may reduce the risk of cancer, heart disease and other chronic diseases.
- Fruits and vegetables also provide essential vitamins and minerals, fiber, and other substances that are important for good health.
- Most fruits and vegetables are naturally low in fat and calories and are filling.[109]

We need plant life to exist! This is pure and simple. Meat eaters get their plants from eating animals that ate plants. Vegetarians get them directly from the plants. Why get them indirectly when you can get them directly? For those that are saying "I am a meat eater and won't give it up" Try to do these things:
1. reduce red meat to a minimal
2. eat lean Organic meats
3. eat wild fish never farm grown (they feed them things that will make you sick)
4. limit your total animal protein to no more than 10% of your daily calories [110] (this includes all animal products consumed such as meats, fish, eggs, yogurt, cheese, milk etc.)
5. increase your daily amounts of fruits and vegetables, mix it up and try new ones and find what you enjoy
6. try to reduce or eliminate milk from your diet (e.g. use soy milk or almond milk)

## Healthy Ethnic Foods

You can enjoy ethnic foods and still eat healthy. Here are 10 ethnic cuisines and examples of what you can order.[111]

1. **Greek**—Greek foods are part of Mediterranean Diet and contain lots of vegetables and dark leafy greens. They also have fruits, beans, lentils, grains and olive oils.
2. **California Fresh**—This is all about seasonal local foods. This includes fruits and vegetables from local farmers.
3. **Vietnamese**—This diet includes fresh vegetables, fruits, herbs and seafood.
4. **Japanese**—Yams and green tea; veggies like bok choy, seaweed, seafood, shiitake and whole-soy foods. Soy dishes such as tofu, edamame, miso, and tempeh, a nutty tasting soybean cake are very healthy and delicious.
5. **Indian**—Use spices that may protect you from certain cancers. Spices such as Curry, Turmeric and Ginger are very healthy.
6. **Italian**—Italian dishes use healthy ingredients such as tomatoes, olive oil, garlic, oregano, parsley, and basil.
7. **Spanish**—Many Spanish restaurants serve Tapas (small dishes) that are a healthy portion. The Spanish also eat plenty of fresh seafood, olive oil and vegetables.
8. **Mexican**—Mexican diet of beans, soups, and tomato-based sauces are healthy. Stay away from fat laden dishes in most American versions of Mexican restaurants.
9. **South American**—They use plenty of fresh fruits, vegetables and seafood.
10. **Thai**—Can a soup fight cancer? If it's a Thai favorite called Tom Yung Gung, the answer just might be yes. Made with shrimp, coriander, lemongrass, ginger, and other herbs and spices used in Thai cooking, the soup was found to possess properties 100 times more effective than other antioxidants in inhibiting cancerous-tumor growth. They use healthy ginger, lemon grass and turmeric in many dishes.

Always eat fruits and vegetables with lots of different colors. Eat nuts and seeds as well. These types of foods make us healthier, protect us from many diseases, improve our sex life and make us look younger.

Monounsaturated fats found in olive oil, fish, nuts, and seeds have been shown to lower the risk of a host of age-related diseases: arthritis, heart disease, diabetes, stroke, cancer, and even Alzheimer's. So if you still have an aversion to the 'F' word, it's time to get over it. Your appearance will benefit, too: The more omega-3's (mostly found in fatty fish like wild salmon) you consume, the more you reduce your risk of age-related skin damage.[112]

Yahoo Diet's article on "The Age Erasing Diets suggests 5 steps to looking younger: [113]
1. Shop for color
2. Fatten up with good fats like Omega-3
3. Sip Red Wine
4. Drink Green Tea
5. Eat less

## The Nine Essential Amino Acids and Where to Find Them

Besides animal based foods (meat, eggs, milk, butter, etc) the following tables shows where we get the essential amino acids from.

| Essential Amino Acids | Foods Containing The Amino Acid | What this amino acid does |
|---|---|---|
| Isoleucine | soy protein, seaweed | Prevents certain generic diseases |
| Leucine | seeds, soy, parsley | Protein synthesis and immune system |
| Lysine | watercress, soybean, spinach, pea, legumes | Oxidation of fatty acids |
| Methionine | sesame seeds, brazil nuts, plant seeds, cereal, legumes, etc. | Fat and protein metabolism |
| Phenylalanine | soy, seeds | Produces important homones |
| Threonine | soy, seeds, tofu, | Breaks down uric acids |
| Tryptophan | soybean, seeds, banana, chocolate (unsweetened), oatmeal, potatoes | brain neurotransmitter that regulates appetite, pain, mood and sleep |

| Valine | soy, seeds, tofu | Growth and maintenance of body tissue |
| --- | --- | --- |
| Histidine* | salad, vegetables. | Promotes growth and repairs the body |

> \* —Essential in infants—after several years of age this is created by the body, so adults effectively have only 8 essential amino acids.

**Figure 13: Table of Nine essential Amino Acids and Foods that Contain Them**

Figure 13: Table of Nine essential Amino Acids and Foods that Contain Them shows the nine essential amino acids (eight for non-infants) and food sources other than meat. It is clear we can easily find the eight essential amino acids we require as an adult with a fruit and vegetable diet.

The ultimate source of the nine amino acids is from plants. Vegetarians should eat a well balanced diet with a variety of fruits and vegetables. Our digestive tracts mix all the amino acids in the foods we eat and we end up with all nine essential amino acids. You don't even need to eat them all at one meal. Our bodies stores amino acids for building proteins later on.[114]

Our bodies break down the proteins we eat into amino acids or chains of amino acids. They are later used by our various cells to rebuild needed protein. Some of this protein will become part of the cells machinery e.g. the cellular walls. Other protein and amino acid chains act as a messenger (epigenetic) to activate or deactivate our genes. We believe that eating living organisms close to us in DNA may trigger genes incorrectly while eating foods farther away on the evolutionary scale are safer. Animals we eat are 85% or higher the same DNA as we are while fruits and vegetables are around 50% the same.

More and more people are discovering vegetarian or near vegetarian diets. They are discovering that there is a life after meat.[115]

## Vegetarians and Protein

As vegetarians, we are often asked in horror, *"how do you get your protein?"* People are dumb struck when they learn that protein is everywhere. Every living thing on Earth is made of protein. This includes fruits, vegetables, fish, poultry and animals. Some scientists argue meat is a more perfect protein because it has all nine essential amino acids. There is disagreement on whether getting the nine essential amino acids in one source (meat) or many sources (a variety of fruits and vegetables) are better. After all, non carnivorous animals get their amino acids from plants.

## Table of Amount of Protein in many Foods

We typically need between 40 and 70 grams of protein per day. Protein is a toxin so eating too much of it places a burden on our kidneys and liver. We need protein but not as much as modern diets provide. The following tables are from buzzle.com

| Food Item | Weight in gram | Protein Content in gram |
|---|---|---|
| Asparagus (boiled, cooked, drained) | 60-62 (4 spears) | 1.55 |
| Artichokes (boiled, cooked, drained, without salt) | 120-122 | 4.18 |
| Avocados (Raw from California) | 28 (1 oz) | 0.60 |
| Alfalfa seeds (in raw and sprouted form) | 33-35 | 1.32 |
| Baked Beans (canned) | 254-255 | 12.17 |
| Broccoli (raw) | 88-90 | 2.62 |
| Beets (cooked, boiled, drained) | 170-172 | 2.86 |
| Bulgur (cooked) | 182-184 | 5.61 |
| Cabbage (raw) | 70-72 | 1.01 |
| Carrots (raw) | 10-12 (1 medium) | 0.08 |
| Cauliflower (raw) | 100-102 | 1.98 |
| Celery (raw) | 120-122 | 0.90 |
| Cucumber (raw and peeled) | 119-120 | 0.68 |
| Dandelion Greens (cooked, boiled, drained, without salt) | 105-107 | 2.10 |
| Endive (raw) | 50-52 | 0.63 |
| Garlic | 3 (1 clove) | 0.19 |
| Lettuce (raw) | 56-57 | 0.73 |
| Lentils (mature seeds, cooked, boiled, drained, without salt) | 198-200 | 17.86 |
| Mushrooms (raw) | 70-72 | 2.03 |

| | | |
|---|---|---|
| Mustard Greens (cooked, boiled, drained without salt) | 140-142 | 3.16 |
| Noodles (chow mein, Chinese) | 45-46 | 3.77 |
| Okra (cooked, boiled, drained, without salt ) | 160-162 | 2.99 |
| Olives (ripe, canned) | 22-23 (5 large) | 0.18 |
| Onions (raw) | 110-112 (1 whole onion) | 1.28 |
| Oat bran (raw) | 94-95 | 16.26 |
| Pumpkin (without salt, canned) | 245-247 | 2.70 |
| Peppers (green, raw, sweet) | 149-150 | 1.33 |
| Peppers (red, raw, sweet) | 119-120 (1 pepper) | 1.06 |
| Potato pancakes (homemade) | 76-77 (1 pancake) | 4.68 |
| Quinoa (cooked) | 1 cup | 11 |
| Radishes (raw) | 4.5-4.6 (1 radish) | 0.03 |
| Spaghetti (cooked, whole wheat) | 140-142 | 7.46 |
| Spaghetti (enriched, cooked, without salt) | 140-142 | 6.68 |
| Spinach (raw) | 30-32 | 0.86 |
| Sweet Potato (canned) | 255-256 | 4.21 |
| Tomatoes (sun-dried) | 2-3 (1 piece) | 0.28 |
| Tempeh | 225-226 | 31 |
| Tofu (firm, made with magnesium chloride and calcium sulfate) | 120-122 (1 piece) | 7.86 |
| Whole wheat bread | 2 slices | 5 |
| Watermelon (raw) | 286-288 (1 wedge) | 1.77 |
| Wheat flour (whole-grain) | 125-127 | 16.44 |

**Figure 14: Table of Protein in Vegetables**

| Type of Bean | Proteins (grams) |
|---|---|
| Azuki Beans | 17.30 |
| Black Beans | 15.24 |
| Broad Beans | 12.92 |
| Cranberry Beans | 16.53 |

| | |
|---|---|
| Garbanzo Beans | 14.53 |
| Great Northern Beans | 14.74 |
| Green Beans | 2 |
| Kidney Beans—All Varieties | 15.35 |
| Lima Beans (Canned) | 10.10 |
| Lima Beans | 11.58 |
| Mung Beans (Canned) | 1.75 |
| Navy Beans | 15.8 |
| Pink Beans | 15.31 |
| Pinto Beans | 15.8 |
| Shellie Beans | 4.31 |

**Figure 15: Table of Proteins in Beans**

It is easy to see that eating a variety of vegetables provides all the protein we need. No one really wants to hurt their kids or grandkids but we must take a strong look at our reward systems and diet. If we don't, they are condemned to a life of suffering and possibly illness. See Chapter 14: How do you get your protein for more information? The following table lists websites that give the amount of protein in grams for popular foods.

| Food | Website |
|---|---|
| Fruit | http://www.highproteinfoods.net/fruit |
| Nuts | http://www.highproteinfoods.net/nuts-seeds |
| Vegetables | http://www.highproteinfoods.net/vegetables |
| List of different food categories | http://www.highproteinfoods.net/ |

**Figure 16: Websites that list protein in different foods**

## Vegetarian Eating

Vegetarian choices are appearing on more and more menus as well as a vast choice of cook books are available to the reader. Vegetarian cooking has truly become a cuisine in its own right.[116] Every cuisine has its own vegetarian recipes. Large cities like New York, Chicago and Los Angeles have an abundance of restaurants that are totally vegetarian (some even vegan) or have a variety of vegetarian and vegan choices on their menus. Smaller cities are starting to change their menus as well.

George used to sell ads for hotels. He got to sample many restaurants on the east coast. He found more and more of them serving veggie burgers or portabella mushrooms as an alternative to burgers. Even Burger King has been offering a veggie burger alternative for years at all of its locations.

George has made chili for his guests made from soy instead of ground beef. No one knew the difference. Only after realizing that George was eating the chili as well, did they begin to suspect something was up. Today there are great non meat alternatives for:

- Burgers
- Meat balls
- Ground beef
- Chili
- Cold cuts
- Meat sauce for pasta (gravy)
- Hot dogs
- German bratwurst
- Spanish Chorizo
- Italian sweet & hot sausage

Try different brands to find which products taste good to you. Not all products are the same. Jo Anne loves Morning Star's Vegan Grillers while George likes Morning Star's Tomato and Basil Pizza Burgers. If you try one brand and dislike it, don't say "I don't like veggie-burgers." Try other brands and other types within a brand. There is enough of a choice these days for you to find one you like.

Typically a great hamburger is not as much about the burger but what you put on it. Typical hamburger toppings are onions, tomatoes, mushrooms, etc. Dress up your veggie-burgers the same way you liked your hamburgers. Sloppy Joes made from non meat substitutes are great. The flavor comes from the sauce not the meat. Cold cuts are the same as burgers. A great sandwich is about what you put on it besides the meat. Dress up your non meat cold cut sandwiches with the same things you would have put on meat cold cuts. Sausage is made with meat fat and spices. The meat fat is not appealing the spices and texture is. Non meat sausages are made with the same spices as the meat alternatives were.

## Health Reasons for Being a Vegetarian

Health is a major reason for becoming a vegetarian. Health reasons include:
- Vegetarians suffer less from many serious and dangerous diseases[117]
- Numerous studies champion the health benefits of vegetarian diet choices by revealing that vegetarians have lower risk for developing the following diseases:
  - Alzheimer's Disease
  - Arthritis
  - Cancer—in particular breast, colorectal, lung, ovarian and prostate cancers. Also
  - Esophageal and liver cancers.
  - Clogged arteries
  - Colon diseases
  - Dementia
  - Diabetes
  - Fibroid uterine tumors
  - Heart diseases, including heart attacks
  - High cholesterol
  - Hypertension (high blood pressure)
  - Kidney disease
  - Obesity
  - Osteoporosis
  - Prostate disease
  - Stroke

It's a very impressive list that makes it worth cutting out all or most meat. Eat your veggies!

## *Advanced*

## Statistics on Vegetarian Health

The following statistics on the affects of vegetarianism on health were found at Dallas—Fort Worth Vegetarian Education Network.[118]

### Cancer[119]

- Breast cancer rate for affluent Japanese women who eat meat daily is 8.5 times greater than poorer Japanese women who rarely or never eat meat.
- a low-fat plant-based diet would not only lower the heart attack rate about 85 percent, but would lower the cancer rate 60 percent.[120]
- 41% reduction in risk of prostate cancer for men whose intake of cruciferous vegetables is high.[121]
- 70 % increase in risk of prostate cancer for men who consume high amounts of dairy products.[122]
- 200% to 300% greater risk of colon cancer for people who eat poultry four times a week compared to those who abstain.[123]
- Vegetarian diets decrease the risk of cancer.
- "Choose predominantly plant-based diets rich in a variety of vegetables and fruits, legumes, and minimally processed starchy staple foods." [124]

### Heart Disease

- Not only is mortality from coronary artery disease lower in vegetarians than in non vegetarians, but vegetarian diets have also been successful in arresting coronary artery disease[125]
- Heart disease can actually be reversed without medicines through a vegetarian diet in which less than 10 percent of calories were contributed by fat combined with a program of modest exercise, no smoking, and stress reduction.[126]

### High Blood Pressure

- The incidence of high blood pressure in meat eaters compared to vegetarians is nearly triple.[127]
- The incidence of very high blood pressure is 13 times higher in meat eaters than vegetarians.[128]

### Obesity/Body Weight

- The obesity rate among the general U.S. population is 18%. The obesity rate among vegetarians is 6%[129]

- 25% of U.S. children are overweight or obese. Only 8% of U.S. vegetarian children are overweight or obese.[130]

## Medical Costs

A minimum of $60,000,000,000 in annual medical costs in the United States is directly attributable to meat consumption. Compare that to $65,000,000,000 in annual medical costs directly attributable to smoking.[131] We all see the risks of smoking but why can't we see the risks of meat consumption?

## Osteoporosis

Current research shows calcium loss to be a larger determining factor in osteoporosis than calcium ingested. Animal proteins, high in sulfur-containing amino acids, especially cystine and methionine, tend to acidify the blood. During the process of neutralizing this acid, bone dissolves into the bloodstream and filters through the kidneys into the urine. Meats and eggs contain two to five times more of these sulfur-containing amino acids than are found in protein from plant foods.[132] Research shows that when animal proteins are eliminated from the diet, calcium losses are cut in half.[133]

The overwhelming evidence is that vegetables and fruits are really good for us just like our moms said they were. Why have we forgotten? Do the math . . . eat less animal based products and more fruits and vegetables and live a healthier life. If you have a chronic illness or are dying, change your diet drastically to a vegan style vegetarian diet. It may help your body combat the disease and cure you. It can't hurt you, that is for sure.

## *Science*

Scientific studies have shown that a lifelong vegetarian diet reduces the risk of colorectal cancer.[134]

"A well-planned vegetarian diet is a healthy way to meet your nutritional needs," said Yogesh M. Shastri, M.D., of Johann Wolfgang Goethe University Hospital, Frankfurt, Germany and previously a co-author of this study at TMH, Mumbai, India. "The exact mechanism by which lifelong vegetarianism may reduce the risk of sporadic CRC needs further investigation. Prolonged vegetarianism starting in early life may be a viable lifestyle option for those at risk of developing the disease."

Science has also shown us that foods that are similar to what was eaten during the Stone Age may help prevent diabetes type 2.[135] "During 2.5 million years of human evolution, before the advent of agriculture, our ancestors were consuming fruit, vegetables, nuts, lean meat and fish. In contrast, cereals, dairy products, refined fat and sugar, which now provide most of the calories for modern humans, have been staple foods for a relatively short time."

"If you want to prevent or treat diabetes type 2, it may be more efficient to avoid some of our modern foods than to count calories or carbohydrate," says Staffan Lindeberg at the Department of Medicine, Lund University.

**History of Vegetarianism**

Vegetarianism began in the Mediterranean and Indian areas around 569-475 BC. India's Buddhists initiated a practice of vegetarianism around the 5th century BC. In Buddhism and Hinduism, a vegetarian diet is still an important practice. Both European and North American cultures had a revival in vegetarianism during the 17th and 18th centuries and even stronger during the 19th century.[136]

## *Conclusion*

Meat may be linked to several of America's chronic illnesses like: 1) heart illness, 2) Diabetes, 3) Cancer and 4) High Blood Pressure.

Eating bacon, sausage, hot dogs and other processed meats can raise the risk of heart disease and diabetes, U.S. researchers said in a study that identifies the real bad boys of the meat counter. This study, an analysis of other research called a meta-analysis, did not look at high blood pressure or cancer, which are also linked with high meat consumption.[137]

None of us want to be sick. No one wants to be in pain or feel helpless. Start thinking, start changing, and enjoy your healthy life. The promise of the drug industry may extend life in the short term, but the life we live will be in pain and full of illnesses. We are not getting healthier as a nation. We are getting sicker! We are taking more and more drugs! We are spending more time with doctors and in hospitals! Is this the life you really want?

Jo Anne & George's typical daily diet consists of:

| Food | Protein (g) |
|---|---|
| **Breakfast** | |
| Oatmeal | 6 |
| Banana | 1.5 |
| Strawberries or other berries | 1 |
| **Snack** | |
| Nuts | 10 |
| **Lunch** | |
| Salad or fruit or vegetables | 5 to 8 |
| **Snack** | |
| Fruit | 2 to 4 |
| **Dinner** | |
| Beans or Soy Product | 30 to 40 |
| Potato | 5 |
| Vegetables | 4 to 6 |
| **Total Protein** | **49.5 to 66.5** |

**Figure 17: Our Typical Daily Protein intake**

As the table shows, we get just the right amount of protein. A 14 oz steak has about 109 g of protein. This is over the limit on one single item. We have seen restaurants offering 24 oz steaks.

Eating healthy as a vegan, vegetarian or eating some limited meat is not difficult. There are plenty of books on why would should do it besides this one and many cookbooks on how to do it. The table below shows some good ones to start with:

| Book, Author and Year | Cookbook or Theory Book | Comments |
|---|---|---|
| **The China Study** T. Colin Campbell, PHD 2005 | Theory Book | Fascinating study of how meat can actually turn ON cancer cells and lack of meat can reverse it by turning them OFF |

| | | |
|---|---|---|
| **The Biology of Believe: Unleashing The Power of Consciousness, Matter and Miracles** <br> Dr Bruce H Lipton <br> 2008 | Theory Book | How our mind and thoughts can affect our health through epigenetics. |
| **The Spectrum** <br> Dean Ornish, MD <br> 2008 | Theory Book and Cookbook | Great way to compare where you are today and where you want to be. Includes a meditation CD |
| **Skinny Bitch—Ultimate Everyday Cookbook** <br> Kim Barnouin <br> 2010 | Cookbook | Great vegan recipes that will make you want to eat healthy |
| **Cooking The Whole foods Way** <br> Christina Pirello <br> 2007 | Cookbook | Great book on eating nutritionally and vegan for anyone |

**Figure 18: Must Read Books & Cookbooks**

A must watch set of videos is by Dr Michael Klaper MD explains how eating animal flesh and fat can make us ill. From 1090 to 1985 the American diet changed by reducing plant based diet in half and doubling animal flesh diets. During this same period Americans became sicker. What you learn can save your life and make you healthier. He states the average American eats 15 cows, 24 hogs, 12 sheep and 900 chickens in their life. Here is a list of the six parts of this video on YouTube:

1. http://www.youtube.com/watch?v=TF2MZN6ImB0
2. http://www.youtube.com/watch?v=Zn-7sfxSmWc
3. http://www.youtube.com/watch?v=41Y2qnE373k
4. http://www.youtube.com/watch?v=dyB79viq74w
5. http://www.youtube.com/watch?v=K4C-2Um4HzM
6. http://www.youtube.com/watch?v=029f9JppOtA

If someone invented a pill that would keep you healthy, make you happy, improve your sex life and decrease your stress, do you think it would sell? That is a stock everyone would want to buy. Guess what? We already have it! It is called fruits and vegetables and it's cheap! Why would we take the artificial pill but not eat the fruits and vegetables. We really are a silly species aren't we?

# Chapter 8:

# Glycemic Index

*Introduction*

America is a nation that worries about sugars. Many people feel fruits are bad because they have sugar. Watch your desserts. As we said before, it's the package that counts. Not the fact sugar is there. Sugary desserts are very bad for us. Adding sugar to foods to addict us is even worse. Fruits are great for us because they are packaged in fiber.

How can we tell what is good sugar and bad? A Glycemic Index was developed to show how each food metabolized. Subjects had their blood taken and base sugar levels measured. They were then fed pure sugar and their blood levels were taken to see how fast the sugar levels went up. This base reading of glucose was arbitrarily set to 100 on the scale and everything else relative to glucose. Levels less than 55 are low and good for you; any level between 55 and 69 are medium and in limited quantities are good for you; and levels of 70 or above are high and should be avoided all together or eaten rarely. Below 69 is what our bodies were used to over the 2 million years we developed. After all what did we mostly eat during that period; fruits, vegetables and some meats? Every item above 69 with the exception of bread, rice, Watermelon and white potatoes were not around even 1,000 years ago. The table below is a Glycemic Index Table showing how each food group rates.

# Table of Glycemic Index and load values[138]

| The average GI of 62 common foods derived from Multiple studies by different laboratories ||
| --- | --- |
| **High-carbohydrate foods** | **GI** |
| White wheat bread* | 75±2 |
| Whole wheat/whole meal bread | 74±2 |
| Specialty grain bread | 53±2 |
| Unleavened wheat bread* | 70±5 |
| Wheat roti | 62±3 |
| Chapatti | 52±4 |
| Corn tortilla | 46±4 |
| White rice, boiled* | 73±4 |
| Brown rice, boiled | 68±4 |
| Barley | 28±2 |
| Sweet corn | 52±5 |
| Spaghetti, white | 49±2 |
| Spaghetti, whole meal | 48±5 |
| Rice noodles † | 53±7 |
| Udon noodles | 55±7 |
| Couscous † | 65±4 |
| **Breakfast Cereals** | |
| Cornflakes | 81±6 |
| Wheat flake biscuits | 69±2 |
| Porridge, rolled oats | 55±2 |
| Instant oat porridge | 79±3 |
| Rice porridge/congee | 78±9 |
| Millet porridge | 67±5 |
| Muesli | 57±2 |
| **Fruit and fruit products** | |
| Apple, raw † | 36±2 |

| | |
|---|---|
| Orange, raw † | 43±3 |
| Banana, raw † | 51±3 |
| Pineapple, raw | 59±8 |
| Mango, raw † | 51±5 |
| Watermelon, raw | 76±4 |
| Dates, raw | 42±4 |
| Peaches, canned † | 43±5 |
| Strawberry jam/jelly | 49±3 |
| Apple juice | 41±2 |
| Orange juice | 50±2 |
| Vegetables | |
| Potato, boiled | 78±4 |
| Potato, instant mashed | 87±3 |
| Potato, French Fries | 63±5 |
| Carrots, boiled | 39±4 |
| Sweet potato, boiled | 63±6 |
| Pumpkin, boiled | 64±7 |
| Plantain/green banana | 55±6 |
| Taro, boiled | 53±2 |
| Vegetable soup | 48±5 |
| Dairy products and alternatives | |
| Milk, full fat | 39±3 |
| Milk, skim | 37±4 |
| Ice cream | 51±3 |
| Yogurt, fruit | 41±2 |
| Soy milk | 34±4 |
| Rice milk | 86±7 |
| Legumes | |
| Chickpeas | 28±9 |
| Kidney beans | 24±4 |
| Lentils | 32±5 |
| Soya beans | 16±1 |

| Snack products | |
|---|---|
| Chocolate | 40±3 |
| Popcorn | 65±5 |
| Potato crisps | 56±3 |
| Soft drink/soda | 59±3 |
| Rice crackers/crisps | 87±2 |
| Sugars | |
| Fructose | 15±4 |
| Sucrose | 65±4 |
| Glucose | 103±3 |
| Honey | 61±3 |
| Data are means. *Low-GI verities were also identified. † Average of all available data. | |

**Figure 19: Glycemic Index Table**

A full listing of foods can be seen at http://www.mendosa.com/gilists.htm

## *Basics*

### Not all sugars are the same

This Glycemic index is important in understanding which foods raise our blood sugar levels slowly (or too quickly). Normally after we eat and begin our routine activity, our blood sugar levels slowly begin to fall as the stored energy is used up. At some point the brain's hypothalamus signals the stomach to give us hunger indicators (growls and pangs). The longer we wait the more intense are these indicators. As we start to eat, blood sugar levels increases, the brain (hypothalamus) stops signals to the stomach that we are hungry. We stop eating and are satisfied. Modern diets often have us eating foods that are full of sugars or break down into sugars with a high GI (see Figure 19: Glycemic Index Table) These quick rises can cause cravings to eat even more. The body tries to respond to the high amounts of sugar with insulin (required for cells to be able to absorb the sugar as energy). If this happens often, we stress the cells of the body causing them to resist the insulin (Type 2 diabetes). Type 1 diabetes is when the pancreas stops producing insulin. In both cases the blood stream is left with an abundance of glucose (sugar). This can lead to vascular problems, nerve problems, and other complications.

Dr Andrew Weil, MD says that a high-Glycemic-index food stresses the pancreas and, in many people, promotes weight gain and unhealthy distribution of fats in the blood and tissues.[139]

Foods that are low on the glycemic index appear to have less of an impact on blood sugar levels after meals. People who eat a lot of low glycemic index foods tend to have lower total body fat levels. High glycemic index foods generally make blood sugar levels higher. People who eat a lot of high glycemic index foods often have higher levels of body fat, as measured by the body mass index (BMI).[140]

## *Advanced*

### Diabetes

Diabetes mellitus, often simply referred to as diabetes—is a condition in which a person has high blood sugar, either because the body doesn't produce enough insulin, or because cells don't respond to the insulin that is produced. This high blood sugar produces the classical symptoms of polyuria (frequent urination), polydipsia (increased thirst) and polyphagia (increased hunger).

There are three main types of diabetes:
- Type 1 diabetes: results from the body's failure to produce insulin, and presently requires the person to inject insulin.
- Type 2 diabetes: results from insulin resistance, a condition in which cells fail to use insulin properly, sometimes combined with an absolute insulin deficiency.
- Gestational diabetes: is when pregnant women, who have never had diabetes before, have a high blood glucose level during pregnancy. It may precede development of type 2 DM.

Other forms of diabetes mellitus include congenital diabetes, which is due to genetic defects of insulin secretion, cystic fibrosis-related diabetes, steroid diabetes induced by high doses of glucocorticoids, and several forms of monogenic diabetes.[141]

## Blood Sugar Levels

### How do we stabilize and lower our blood sugar through diet?[142]
1. Avoid eating foods with high sugar content. Foods high in sugar cause our blood sugar levels to sky rocket.
2. Eat several small meals throughout the day. Large meals spike blood sugar and insulin production. Eating more but smaller meals allows our blood sugar levels to remain more stable.
3. Don't eat carbohydrate-only meals. Healthy fats like monounsaturated fat (found in olive oil) and omega 3 polyunsaturated fats (found in salmon and flax seed oil) help to slow down the release of carbohydrates (sugars) into the blood stream. The slow release allows for lower steadier blood sugar levels. Proteins also help to slow down the release of carbohydrates into the blood stream. Eating a banana, drinking a glass of orange juice, or having a few crackers on an empty stomach is a major no-no. This type of eating will cause your blood sugar and insulin to spike.
4. Choose complex carbohydrates over simple carbohydrates. Complex carbohydrates break down much more slowly and have a higher nutritional value than simple carbohydrates. Complex carbohydrate sources are whole grains, most vegetables, brown rice, unprocessed oatmeal, and whole grain pasta. Simple carbohydrates include pop, junk food, white rice, bread, bananas, cereal . . . .
5. Eat a high fiber diet. Foods that is high in fiber help to slow down the release of sugars into the bloodstream.
6. Don't eat too much. Overeating is a major problem especially in the developed world. These extra calories cause excess insulin production as well as increased insulin resistance.

We believe bananas are fine for healthy eating but if you have diabetes you may want to limit or eliminate them. If you want to prevent diabetes or if you have it and want to cure it; change your life style and diet. Don't wait until you are pre-diabetic, act now.

## *Science*

Glycemic Index (GI) is a ranking system for carbohydrates based on their immediate effect on blood glucose levels.[143] Carbohydrates that break

down rapidly in our bodies have the highest GI. Carbohydrates that break down more slowly, release sugar into our system slower and have a lower GI. It is not just about sugars but again how they are packaged that affects our bodies. High GI carbohydrates can cause an increased risk in:
- Woman for heart disease
- Development of type 2 diabetes
- Increased weight gain

## *Conclusion*

Diabetes is said to be incurable. Once you have it you have it for life however, lowering your blood sugar levels through diet will cause the symptoms to disappear and the disease will be under control. This will greatly help eliminate any further body damage done by the disease. Diabetes can be cured by epigenetics. Changing your diet, changes your epigenetics. Stop feeling doomed and reset your life style. It can't hurt and probably will help you get back to a healthy lifestyle.

# Chapter 9:
# Take Control of your Thoughts, Actions and Diet

## Introduction

In the three environment chapters on food, thought and belief we learned that each of these can affect our epigenetics in negative ways and lead to diseases. In this chapter we will cover why this happens so often and what we can do to control each of these.

## Basics

### Mass Misinformation

We have mentioned that the power of marketing is to sell us stuff even if we don't need or want it. It can also be used to build a set of basic beliefs that become doctrines to the general public. These new doctrines become entrenched in our brains and get passed along to new generations causing us to believe in things that are not true. Here are some examples of mass generated misinformation:

1. ***We have no control over our lives; it is controlled by our DNA.*** If we get a good set of DNA we will be healthy, if not we will be sick. This was believed even by many scientists during the last decade. Evolution seemed to be solely random changes to our DNA. We now know that there is something operating above and on our DNA called epigenetics. Epigenetics has the ability to

react to our environment (food, temperature, toxins, stress, etc) much faster than mutations to DNA. In fact some of these changes are immediate. The changes to our epigenetics causes us to stop "playing" some genes and start "playing" others. Sometimes these changes are good and allow our species to survive drastic changes to our environment but sometimes that cause negative effects on us like heart disease, cancer and diabetes. **ACTION: Instead of giving up on your DNA, train yourself to make positive changes in your life and allow your epigenetics to "play" healthy genes for you. Our bodies are very powerful if given the right stimulus. Just like your car won't run very well or long if you feed it garbage fuel, our bodies need healthy fuel to survive. Many people are more concerned about what they feed their cars than what they feed themselves and their families. Are you like this?**

2. *"Bet you can't eat just one!"* worked for Lay's because it caused us to believe it and buy more and eat more chips. Since this food gives us a brain pleasure rush we want more and more of them. The problem is they have little or no value as a fuel to our body. Is your body screaming at you to feed it something healthy? It may be trying to tell you by causing disease, obesity, pain, etc. Are you listening to your body? **ACTION: Try eating a healthy diet for 2 months. One primarily of fruits and vegetables. Try using some raw vegetables as well. NOTHING from a can, box, or package. Stop junk foods during this time. Stop eating out as much since portion sizes are usually large. You will save money and help your body return to a healthy state. It's only 2 months' you can do it. You will find the fruits and vegetables taste so much better to you. You will start to lose weight. You will feel better with more energy. Go ahead give it a try. You can only win. After all no one will say eating fruit and vegetables is bad for you.**

3. *Anything in moderation is OK for you.* It is not. Try eating poison in moderation. Oh you are aren't you? Junk food and fried foods and too much animal protein are poison. It is toxic and will cause your body to age, gain weight and be more prone to chronic illnesses. It will also kill your sex drive! **ACTION: Write down what you eat for a week. List everything. You don't have to tell anyone or show anyone else. You don't have count calories. This is an exercise for you alone. List next to each item, who made it. Next to natural foods you list God or Nature. Packaged foods list the**

manufacturer. How many foods are made by God? Probably a painful few. This is why we are sick and getting sicker. If this doesn't bother you, continue your life style. It is, after all, your free will and choice. If on the other hand you are not happy with the direction you are headed, change your direction and reap the positive benefits.

4. *There's a full serving of vegetables in every serving.* If you really believe this I have a bridge for sale in Brooklyn. Having tomato paste in a Sloppy Joe product does not make for a serving of vegetables. Processing food and removing fiber changes the value of the food nature provides us. Don't be fooled by commercials telling you what you can and should eat. They are doing this to make more money, not make you healthier. Ask yourself two questions: A) How long has God/Nature been making fresh fruits and vegetables? B) How long has this manufacturer been making their products? **ACTION: Don't get up and walk away during commercials. List them on a paper with topic, manufacturer, and message they are trying to ingrain in you. Add up how many commercials you saw. How many were for drugs? How many were for manufactured foods? Can you see through each message and identify that only the manufacturer wins, you don't.**

5. *Drugs cure diseases.* Drugs are not designed to cure, but to provide relief from a symptom of the disease for the remainder of your lifetime. Imagine I have a company that sells drugs. I can provide a pill that will cure cancer or AIDS or I can provide a pill you must take for the rest of your life that will give you relief from the disease. Which will I do? The latter of course it is highly profitable and I have a hooked consumer for life. The former will put me out of business after I cure the disease. **ACTION: Listen carefully to drug ads. What are they really saying? How many adverse effects are there? Do they claim to cure anything? Don't you wish you had a product that you could sell to more than half of the Americans for life and it really didn't work?**

6. *Diets and diet companies work.* If they worked, we wouldn't have obesity and they would be out of business. Do they want to be out of business? NO! Diets fundamentally do not work because they turn on our body's starvation mode making it really hard to lose fat. We will lose some weight in water and even some in muscle which is really bad. New diet companies that provide the food are

highly profitable. They show you desserts you want as acceptable foods on the diet. These foods have sugar or sugar substitutes that trigger the brains pleasure center. Eventually your desire wins and you binge. Then after a time you get more frustrated and try it all over again. This viscous cycle makes a lot of money for the diet industry but doesn't work well for us. If it did work, obesity would be on the fall, but it's on the rise. **ACTION: List how many times you went to a company or particular diet and how long it lasted. Ask your friends how many times they did it. If any friends are still obese, it obviously has not worked. Research how many customers these companies have. Research the rate of obesity in America. The truth will set you free!**

7. *We are winning the battle against chronic illnesses like cancer, heart failure, diabetes, Parkinson's, Dementia, Alzheimer's, etc.* This is obviously false. We are failing miserably. More and more Americans are getting sick. We are building more and more hospitals and nursing homes. Drug companies are getting bigger and bigger. Just look around you at all the new CVS, Walgreens, and Rite Aids that are opening. Quality of life is not only way down; no one expects a good quality of life when they get old. This is not the outcome nature or God had set for us. Bad things do happen. We can be involved in a bad accident. We may need surgery or drugs to help in this case, but the vast majority of drug use, hospital cases and people in nursing homes are the result of bad eating, smoking, lack of exercise and stress. **ACTION: Just take your life into your own hands. We are not telling you to stop seeing your doctor or to stop taking your drugs. In fact many drugs can have a very bad affect on you if you stop quickly. Try changing your life. What is the worst thing that might happen; you might get healthier? There is no down side to this. You will not get sicker because you started eating healthier or removed some stress from your life or stopped smoking.**

## *Conclusion*

Good marketing companies get paid the big bucks to fool us. If they can sell you a bad product or negative idea and make it seem like it is good for you, they have done their jobs well. The worst part is most Americans sit in front of the TV all evening. They are exposed to more and more commercials. The concepts in these commercials become facts. If you hear something over and over again, you believe it. The Internet is even worse. Pseudo-facts are published all the time. If we read it online we tend to believe it. Do your own research! Draw your own conclusions! Live YOUR life well!

Finally stop reading papers and watching the news so much! Every story is negative. Negativity sells. We are a morbid race. How often do you see positive things on the news or on the front page of a newspaper? We are members of the Rotary and have seen many positive things that good people do for others. Never does it get published on major networks or newspapers. If you watch the news in the morning and evening and read a paper or two, after a short while, you will believe everything in life is negative. This will put stress on you. How can you cope with this? STOP watching it! Seek out different news institutions to get all points of view. Spend as much time reading about positive stories as you do negative ones. Instantaneous world news is a great technological invention but our bodies haven't evolved to handle the extra stress yet.

# Chapter 10:

# Don't let TV Advertisements Control your Actions

## *Introduction*

Americans consume TV like they do sugar. Marketing companies know this. They target us with ads that we believe because we have heard them so often. What you hear in an ad is not necessarily for your benefit. It is to sell products and make money. Lots of money is made through the use of clever ads.

## *Basics*

### Marketing

Marketing is designed to make you buy something even if you don't want it or need. Try counting the number of TV ads for drugs. It's huge! The revenues from these ads cause newspapers, TV and radio to not want to be "anti drug". Doing so would cause the drug company advertisers to pull their ads. This means we are seeing only one side of this argument. The counter point side never gets stated. Have you ever seen an ad for eating healthy that is not related to a pill or product? How about an ad for meditation? The reason we don't see these is no one makes any money by keeping us healthy, thin and happy.

Imagine an America that eats right, feels good and is not obese. The health care system with all our hospitals and doctors would collapse. No need for the number we currently have. The drug companies would go out of business. The diet companies would as well. Our taxes to cover the huge cost of America's illnesses would decrease drastically. Our politicians would have to find something else to spend their time on. This is not a pipe dream. We can obtain this but we won't if we continue on our current road.

We are on a downward spiral in this country. To cover up the obvious, we are being told it is normal. It is a part of growing old. Bull Sh**! Don't be fooled. Growing old has nothing to do with having a poor quality of life. We make our lives a poorer quality by believing we can't do anything about it.

**Who gains from the ad?**

We have talked about this a lot already. Commercials are for one and only one purpose, to sell you something you may or may not want or need. If the product is not the best for us, it is sold in a light to make it look good. Humor plays a great part here. If they can make you laugh, you will remember the advertisement. Mentally you will have a hard time opposing something you find funny. The humor is an attempt to move our attention off what is being sold and make us laugh and feel good. This causes us to buy the product based on false humor.

The exception to this rule is drug advertisement. Generally illness is serious business and people that are suffering want their conditions taken seriously. Humor would not work well here. There is another powerful motivator at work, if you are sick you want to be cured or at least have the symptoms relieved.

The problem is that this claim doesn't often work. Just look at hospitals and nursing homes. Many people there are not happy. They are suffering and their quality of life is very low. But stay and watch for a while and you will see one thing in common. They all take medication . . . a lot of it.

If we look closer at these institutions we see that there is little quality of food being served. Most is mass produced to save money. Very little fresh product is consumed. Fruits often are in sugary syrup and vegetables have been cooked so much there are no nutrients or benefits to eating them.

George was with his mom at a hospital where she had just had a knee replacement done. She was in a small cafeteria on the floor where most people had their knees replaced. Most were also obese; which is why they needed the surgery in the first place. He looked over the menu for her. It was fried hamburgers, cheese burgers, French fries, ice cream, etc. I asked a nurse why there wasn't anything healthier on the menu. Her response was "These people aren't sick. They can eat what they want." Yes, THEY WERE SICK. They were obese because of their eating habits and that lead to a knee replacement. Feed them what they want so they can get the other knee replaced.

**Follow the Money Trail**

Money is a strong motivating factor. There is a lot of money to be made in diet and medications. More money than most small countries have. The promise is always enticing:
- We will make you healthier
- We will make you thinner
- We will make you stronger
- We will cure your diseases

Who wouldn't want this? The problem is that none of these are happening on a consistent basis to most of the population. The facts are:
- We are getting sicker as a nation (see Chapter 11: America is Sick, Literally below)
- We are an obese nation
- We are becoming weaker
- Very few diseases are ever really cured

So why are so many companies in these businesses? There is a lot of money to be made. There is tons of it in fact. Consumers are not thinking straight. Government is being lobbied to not protect us. No one is watching or really regulating these industries. They are virtually free to do as they please. This abuse is so much bigger than the tobacco company abuse, the Enron abuse or the financial institutions abuse. They all have one thing in common . . . GREED!

## *Conclusion*

Don't be fooled by clever ads. Do your homework and research what the facts really are. You are the only person that really cares about health. We all have one life to live. When it's over, it's gone forever. Lying on your deathbed, sick, is too late to think about what you should have done. Start today and be happy. If you think, we are not really paying for all this, think again. Each and every one of us pays in taxes for this huge health industry burden.

# Chapter 11:
# America is Sick, Literally

## *Introduction*

As a nation, we are so sick; we actually believe its part of growing old. Sickness has nothing to do with being old. Our bodies get tired of a life time of abuse and fail us after time. We are amazed when we hear stories of people over 100 that are strong, healthy, exercise each day. Why? They kept themselves healthy by their healthy lifestyles!

## *Basics*

Our way of life is being jeopardized! We believe we have a great health care system, when in fact we don't. Most doctors, not all, look at your symptoms and prescribe a drug or procedure. This has become health care. It wouldn't be bad if it worked but it is NOT working. We are getting sicker, dealing with chronic illnesses in old age and our youth are beginning to become obese and sick at younger ages. This doesn't have to happen. Dr John Kelly, MD and President of ACLM (The American College of Lifestyle Medicine) says, "Why would we do medicine and NOT treat causes? We only treat symptoms and make diseases livable while the patient continues to die![144]

If you started using some cheap dirty fuel in your car, what would happen? Over time it would start to fail you. It wouldn't last as long as it should. Using the correct fuel and oil will make a car operate better and last longer. Humans are the same. If we use the correct fuel, we will live a healthier life. We all know this deep down.

## *Advanced*

Illness and depression may all be linked to epigenetic changes. Research is showing that each may be caused by a change in our epigenome causing the malady. Since epigenetics is influenced by diet, stress, belief and environment; we can make changes to lower the probability of these maladies. Lifestyle theme[145] published the following list of statistics on chronic disease in America:

- Nearly 1 in 2 Americans (133 million) has a chronic condition. Chronic Care in America: A 21st Century Challenge, a study of the Robert Wood Johnson Foundation & Partnership for Solutions: Johns Hopkins University, Baltimore, MD for the Robert Wood Johnson Foundation (September 2004 Update). "Chronic Conditions: Making the Case for Ongoing Care".
- By 2020, about 157 million Americans will be afflicted by chronic illnesses. According to the U.S. Department of Health and Human Services.—Chronic Care in America
- That number is projected to increase by more than one percent per year by 2030, resulting in an estimated chronically ill population of 171 million.—Chronic Care in America
- 96% of them live with an illness that is invisible. These people do not use a cane or any assistive device and may look perfectly healthy.—2002 US Census Bureau
- Sixty percent of the chronically ill are between the ages of 18 and 64.—Chronic Care in America
- 90% of seniors have at least one chronic disease and 77% have two or more chronic diseases. The Growing Burden of Chronic Disease in American, Public Heal Reports / May-June 2004/ Volume 119, Gerard Anderson, PhD
- 9 million people are cancer survivors with various side effects from treatment.—American Cancer Society
- The divorce rate among the chronically ill is over 75 percent. National Health Interview Survey
- Depression is 15-20% higher for the chronically ill than for the average person. Rifkin, A. "Depression in Physically Ill Patients," Postgraduate Medicine (9-92) 147-154.
- Various studies have reported that physical illness or uncontrollable physical pain is major factors in up to 70% of suicides. Mackenzie

- TB, Popkin MK: "Suicide in the medical patient.". Intl J Psych in Med 17:3-22, 1987
- About one in four adults suffer from a diagnosable mental disorder in a given year. Kessler RC, Chiu WT, Demler O, Walters EE. Prevalence, severity, and comorbidity of twelve—month DSM-IV disorders in the National Comorbidity Survey Replication (NCS-R). Archives of General Psychiatry, 2005 Jun; 62(6):617-27.
- More than 90 percent of people who kill themselves have a diagnosable mental disorder. Conwell Y, Brent D. Suicide and aging I: patterns of psychiatric diagnosis. International Psychogeriatrics, 1995; 7(2): 149-64.
- Four in five health care dollars (78%) are spent on behalf of people with chronic conditions. The Growing Burden of Chronic Disease in American, Public Health Reports, May June 2004 Volume 119 Gerard Anderson, PhD

**Our Government's Impact on Cancer**

"G. Edward Griffin marshals the evidence that cancer is a deficiency disease—like scurvy or pellagra—aggravated by the lack of an essential food compound in modern man's diet. That substance is vitamin B17. In its concentrated and purified form developed specifically for cancer therapy, it is known as Laetrile, the controversial chemical that currently is banned in the United States.

This story is not approved by orthodox medicine. The FDA, AMA, and American Cancer Society have labeled it fraud and quackery. Yet the evidence is clear that here, at last, is one answer to the cancer riddle."[146]

Even though concentrated amounts of **B17** are band in this country, it can be gotten naturally from many sources:
- Many whole raw foods
- Seeds of non-citrus fruits
- Apple, peach, cherry, orange, plums, nectarine and apricot seeds in particular lima beans, clover and sorghum

Americans have the most advanced medical scanners in the world. The number of tests based on radiation is increasing. It is estimated that our dose of radiation has increased 6 fold do to these scans.[147] This radiation raises the risk of cancer. We are all worried about radiation from cell phones, microwaves and airport scanners but these medical scanners are far more powerful. Isn't it strange that we live in a society that scans for cancer using machines that can cause it?

# *Conclusion*

**Do your own research**

It is very clear that we are a nation of sick people and we are getting sicker. Modern medicine is unable to stop it or even slow it down. For all the drugs we take, nothing is working. Do your own research. Think about what is really happening with health in America. Be one to stand up and stand out as an individual who will not allow this to happen to you. Change your life; you only have one shot at it.

**Develop a Plan for your Happiness, Health and Improve your Quality of Life**

Honestly evaluate your life style. Track what you eat for a few weeks. Are you eating fruits and vegetables? Replace your Orange Juice in the morning with a fresh organic orange. Try different fruits and vegetables. You may not like everything you try but you will find some new ones you enjoy. Try BBQ'ing your vegetables for a different flavor and grill experience. Replace some of the meat you are eating with fish. Be sure to select organic wild fish. Evaluate how many sweets you are consuming and try to eliminate most. Replace your oils with extra virgin olive oil. Don't cook or heat your olive oil too much. Try using a spray olive oil on vegetables to minimize usage and then sprinkle a small amount of raw olive oil first cold press. This oil is more expensive so don't cook with it. Just use it on salads and vegetables to add some flavor. If you like fried foods, try replacing them with oven baked or BBQ'ed.

Change your diet little by little to a healthier one. Enjoy your food more and live a better life. Try taking a meditation class to reduce stress on your body. You will sleep better and feel better. Get exercise even if you just start walking. Get out and move your limbs before they won't move anymore.

# Chapter 12:

# Drugs—Pro & Con

## Introduction

We have become a drug nation. If you believe the commercials, pills will cure all. Some of us need some pills to fight an infection or a disease, but many pills are unnecessary. Many people end up taking pills to offset affects of other pills.

## Basics

### Our Environment

We live in a PILL society. Take a pill for anything and everything. The problem is the pills don't really work. If they did we would be getting healthier as a society, not sicker. George often says to his mom, it is not about dying, we are all going to die someday. It's about how we live the last 10 or 20 years of our life that really counts. This is our quality of life. Our society with its pills is causing us to live a little longer with our diseases but our quality of life is poor.

Americans want a quick fix for everything. Take a pill to fix all your ailments, go on a diet or take a pill to lose weight. Take a pill to feel happier. The problem is that these actions are fixing nothing. We are getting fatter, feeling more depressed and getting sicker. The only thing that these actions do is to make certain individuals and corporations very wealthy.

We are sure you know someone that is sick, unhappy or obese. What have they done about it? How long have they been doing it? Are they getting better? The answers to these questions in 99% of the cases are:
1. **What have they done about it?**
    a. *Sick*—They are seeing more and more doctors and taking more and more pills but the illness is still there. Nothing is changing.
    b. *Unhappy*—They are seeing a psychiatrist and maybe taking pills but never really feel great
    c. *Obese*—They have been on one diet or another, have taken pills and maybe even operations to cut out the fat.
2. **How long have they been doing it?**
    a. *Sick*—They got the disease and have been fighting it in some cases for years. Most people eventually die of a serious chronic illness.
    b. *Unhappy*—They have been seeing a Psychiatrist for years and have been on medication for as long, but they still feel depressed.
    c. *Obese*—For some people they have been fighting obesity all their lives.
3. **Are they getting better?**
    a. In most cases NO!

How can we say we are improving? Is this any way to live? Can we do anything about it? We believe the answer is yes, and our ancestors knew it all along! Eat healthy and feel good and be healthy. Don't let pills and stress become a constant attack on our bodies. Instead of putting chemical compounds (pills); which can be toxic and have many side effects; into your body, why not try eating healthy and nutritional foods, thus reducing the need for pills? Find ways to relieve the stress through meditation and exercise.

What is amazing to us is that most scientists who write on epigenetics predict new epigenetic drugs in the future. Why not predict that we will find the true cause of illness like food and thought and help people change it. We as a society are so pill driven we can't think any other way.

## Vitamins and Drug

Real foods beat out pills!

"Following the right diet can literally add years to your life," says Ronald Klatz, MD and Robert Goldman, MD in their book, stopping the clock[148].

We are fast becoming a society of drug peddlers and takers. Some drugs are necessary but most are marketed not to help the patients but to make money for the drug companies. Our comedians even joke about the side effects of some drugs compared to the remedies. George has seen his mom taking a huge amount of drugs some even to counter the effects of others. Whenever she was hospitalized, drugs became a problem causing unneeded aggravation and pain. No one could keep track of the interactions between the drugs.

We are not against the medical community. We have some of the best doctors in the world, but they are not trained in dietary needs and have been trained in school that the solution to all medical problems is to identify symptoms, isolate possible causes and prescribe drugs or surgery. Sure many surgeries are needed but all drugs are not. High blood pressure, diabetes, cholesterol, and cancer are mostly caused by our food, stress and environment. Once we get one or more of these we can either take drugs forever or change our lifestyle and solve the problem. Even young nurses in nursing school are being taught the drug paradigm. Listen to them talk. It is about a symptom and a drug. What a waste of talent. Why aren't we teaching young doctors and nurses about diet? They know about certain foods you should not eat after you have a problem. Why not before having a problem to prevent it?

George, when he was President and Co-CEO of a public company, had a lot of stress on him. During an annual physical, he was told by his doctor that his blood pressure was up and so was his cholesterol. He was prescribed drugs for both. George declined the drugs saying he wanted to research the problem and find alternate solutions. He said he would be back next year and if they were still elevated, he would take the drugs. George decided to change his diet and lifestyle drastically. He joined a gym to get more exercise and lowered his stress and became a vegetarian to eliminate the animal fats from his diet and started to eat healthier. The following year he was told by the same doctor that both his pressure and cholesterol were now low. The doctor wondered how he managed to do that. We do have alternatives and the science of epigenetics is showing us why these alternatives are working. Remember the saying "no man is an island?" Epigenetics is based on this. We humans are constantly interfacing with our environment and food is part of our environment. Each interaction causes small changes in which genes are expressed or prevented from being expressed. These changes can cause us to be healthy and happy or unhealthy and unhappy. The science of understanding all these interactions and what eventually will be affected is very complex. Thank God we don't need to understand any of this. Early man certainly did not. All we have to do is to eat healthier and reduce our stress.

## *Advanced*

### Drug Companies

Drugs are a very big business both the legal and illegal types. Drug companies control which projects get researched. The funded projects are those that will make these companies even more money. Drug Companies fund major medical schools and have powerful allies in the government. Natural products despite their effectiveness are ignored; no money can be made here. The FDA is supposed to be our watch dog in the government but they mostly act as safeguards for the Drug Companies financial interests. Staff from the major drug companies and the FDA staff constantly rotate. [149]

It is not within the drug company's interest to cure anything. They are corporations and as such their mission is to make a lot of money for themselves and their shareholders. This is just how capitalism works. The good news is we do not have to take drugs for everything. We can choose to find alternative ways to gain a healthy life style. We would not care as much if, as a result of these actions, America was getting healthier, but it's not (see Chapter 11: America is Sick, Literally). If you pay that much for a product or service, it should at least work!

### Drug Research

Because drug companies insist as a condition of providing funding that they be intimately involved in all aspects of the research they sponsor, they can easily introduce bias in order to make their drugs look better and safer than they are. Before the 1980s, they generally gave faculty investigators total responsibility for the conduct of the work, but now company employees or their agents often design the studies, perform the analysis, write the papers, and decide whether and in what form to publish the results. Sometimes the medical faculty who serve as investigators are little more than hired hands, supplying patients and collecting data according to instructions from the company.[150]

Drugs are both powerful and dangerous. Since drug companies control the media, the medical schools and the tests that are done, you have a major conflict of interest. Drug companies control the advertisements you see on TV and newspapers and magazines. They pay so much in advertising dollars that they can prevent a negative drug view from happening. This prevents talk shows and news broadcasts from discussing the down side of drugs or alternatives to drugs. They control medical schools by making large donations and funding research programs. This gives them the right to push a drug centric medical viewpoint. It also controls the research to ensure a positive spin on all drugs. They can even use lobbing to control our government. Who is really watching them? It seems to us this industry is totally out of control.

## Medical Schools

Since medical schools are funded by the drug companies they teach the drug paradigm. Most doctors have this ingrained into them by the end of medical school. Then if they are associated with a hospital that gets funded by the drug companies, they must follow suit. To speak against it would be career suicide. There are some smart younger doctors that are resisting. They see that the current strategies are not working. When looking for a doctor, seek out one that shares your beliefs on food and thought. Talk about your medicines; see if your doctor will recommend another approach. If you have researched an alternative approach, mention it to your doctor. If you don't take your health and happiness into control, no one else will.

Drug companies try to buy doctors with gifts of free lunches and other gifts given out by the pharmaceutical industry. Some doctors are saying enough![151] Have you ever visited a doctor that told you to take more fruits & vegetables rather than drugs? There are some, but sadly they are few and far between.

## Drug Advertisements

Advertisements are very powerful motivators to buy something. Drug companies know this too well. That is why we have so many drug related commercials on TV. Have you ever seen a commercial that tells you to eat your fruit and vegetables to stay healthy? Have you seen one that tells you to take walks to get good exercise and lose weight? You never will because no one makes money if you eat fruit & vegetables and take walks.

## Way of Life

Each of us needs to make these changes a way of life for us. It is always interesting to us that when we are in tune with our diets, things like fruits and vegetables taste so good to us. When we eat crap they don't. We develop a taste for what we eat. Eat fatty foods or sugary foods and you will crave more of them. Eat healthy vegetables and fruits and you will crave more of them. Remember **you are what you eat!** Start eating right, in a few days you will be enjoying your food. Experiment with new fresh items.

Eat things raw to preserve the healthy protein in them. When you cook things like vegetables, cook them only a little. Al dente vegetables have a crunch to them; they taste better, have more fiber and nutrition in them. If you are not imaginative, buy a few cook books that cook this way. Get ideas and try them. You will soon find out there are so many things you are enjoying and you will want to cook for your family and friends.

## The Problem

Science observes symptoms and makes up theories to explain these symptoms. If the theories are correct, they will be able to correctly predict future events and become part of the science foundation.

As we better understand the symptoms and proposed theories, we begin to understand the underlining causes of the symptoms. This enriches our understanding of all that is around us. It is this scientific theory that keeps us honest and increases our understanding of God's Universe. The drug companies don't really care about the underlying causes or problems, only the symptoms. If they can help resolve a symptom with a drug, they can make people feel better while *hooking* them on the drug. They will never be cured since they never solved the underlying problem. This way of thinking makes most of us customers and brings unimaginable wealth to the drug companies and their executives. The number of *incurable diseases* is increasing. Isn't it convenient that an incurable disease will get a drug company paid for life? A curable disease doesn't pay much.

Look beyond the drugs to the real solutions. Find doctors that want to help cure you, not just prescribe a drug that solves the symptoms. Find a way to exercise more, eat better foods (more plants), eat less processed foods, Do yoga and meditate to help you sooth the mind and live in as clean a place as possible. These things cost you little or nothing to do and will help make you healthy and happy in life. Remember it is not about dying, we will all die. It's about the quality life we have during our last 10 or 20 years.

Polio was the last disease cured or eradicated in the US. Why? Drug companies no longer want a cure, only a way to better live with the disease. Curing makes no money! Prolonging life and living with it makes a boat load of money. Those are the facts. They aren't pretty but they are true.

## *Conclusion*

Some drugs are necessary to help us fight an infection or disease but many drugs are frivolous and are intended to make a lot of money for the drug companies. Do not stop your medication on your own! Find a doctor that is open to finding other non drug ways to help you. Under the doctor's care, reduce your drug dependency where possible. It will require exercise and an improved diet. If you really don't care about being healthy and happy, no one else will either.

# Chapter 13:
# Sugar addiction

## Introduction

Although scientists do not agree on the addictive powers of sugar, more evidence is showing up indicating it is very addictive. The reviewed evidence supports the theory that, in some circumstances, intermittent access to sugar can lead to behavior and neurochemical changes that resemble the effects of a substance of abuse. According to the evidence in rats, intermittent access to sugar and food is capable of producing a "dependency".[152] Sugar has no redeeming benefits to our life or health. It does give a rush to the brain that we feel as pleasurable. That pleasure makes the brain want more of it. This cycle is an addiction. Sugar is chemically needed by the body to signal events such as appetite and need for release of adrenalin. In foods like fruits and vegetables, it is packaged in a way that our bodies can better respond to. The packaging causes it to be assimilated slower into the blood stream. Pure sugar and man-made sugary products are not packaged this way and enter the blood rapidly, causing problems. In Chapter 8: Glycemic Index, we spoke about measuring how fast sugars enter the body in the Glycemic Index Chart.

## Basics

### Is Sugar an Addiction?

Sugar is the most common addiction of our society.[153] Cynthia Perkins, in her book "*The Hidden Dangers of Sugar Addiction*", says the average

American consumes 32 teaspoons of sugar a day. She says that refined sugar is very different than raw sugars found in fruits and whole foods. She also says: "The list of health problems associated with sugar is enormous and too large to go into completely in this book, but some of the most common problems include:"

- depression
- mood swings
- irritability
- depletion of mineral levels
- hyperactivity
- anxiety or panic attacks
- chromium deficiency
- depletion of the adrenal glands
- type 2 diabetes
- hypoglycemia
- obesity
- Candida overgrowth (yeast infection of the GI tract)
- high cholesterol
- anti-social behavior such as that found in crime and delinquency
- anger control issues
- insomnia
- decreased immune function
- aggression
- neurotransmitter deficiencies
- high blood pressure heart disease asthma
- alcoholism
- acne
- PMS
- OCD
- fibromyalgia
- attention deficit
- cancer
- chronic fatigue
- addiction
- hormone imbalance

As we review the above list we see many conditions and ailments that afflict our society. The diseases alone are concerning. Imagine all of the illnesses we consider to be devastating can be tied back to sugar. In addition, alcoholism, PMS, depression, high blood pressure, aggression, insomnia and mood swings. Does this sound like the type of person you want to be? We all already knew it caused obesity. How about decreased immune function? Doesn't that sound like it could lead to other serious and chronic problems as well?

The consumption of sugar is considered to be one of the three major causes of degenerative disease in America by the American Diabetes Association. Sugar is so destructive it can probably be linked to just about any health condition you can think of, and then some. Cynthia also recommends removing fruit juice from your diet. Remember fruit juice is NOT fruit! Man took the nature made package away and gave us sweet juice in a bottle instead.

Dr. Serge Ahmed, of Bordeaux, France, has been working with rats. He has been doing experiments with the choice between sugar and cocaine. Each time the rats choose sugar over cocaine. The sweet reward of sugar wins over the high that cocaine produces. That would seem very additive to us.[154] Professor Bart Hoebel and his teams at the Department of Psychology and the Princeton Neuroscience Institute are performing experiments that capture the phases of craving and relapse, which complete the addiction cycle.[155] No matter what you have been told; no matter what the evening news says or commercials allude to, SUGAR IS VERY ADDICTIVE AND EXTREMELY DANGEROUS TO LIFE ON EARTH!

All scientists may not agree, but the experiments do indicate that sugar is our enemy. We must find a way to control it in our lives and eliminate as much of it as possible. Just like salt, the amounts in processed foods are a real killer. They seem to addict us to the foods and cause grave health problems. Most of us would not want the 32 tsp of daily sugar spoken about above but we get it hidden in the various foods we eat.

## *Advanced*

### Adding Sugar to foods

Why is sugar added to food and drinks? While added sugar provides no nutritional value, it does serve many uses in food processing. Added sugar can:[156]

- Serve as a preservative for jellies and jams
- Provide bulk to ice creams
- Assist in fermentation of breads and alcohol

- Maintain the freshness of baked goods
- Sugar is also added to processed food and drinks because it makes them taste more appealing.

Adding sugar to unlikely foods such as milk causes kids to like milk even if they don't like the basic taste of milk. [157]

Do we need to have an addictive substance added to our foods to like them? Are you starting to see the differences between artificial commercial foods and nature provided real ones? Nature has packaged food to provide us with everything we need to be happy and healthy. The packaging was worked out over millions of years. We evolved with these food supplies. They are natural to us. Commercial foods are at best a few hundred years old and the most destructive are only about 50 years old. The purpose is not to nourish us but to addict us and make money from us. Are you paying to be unhealthy?

Imagine if someone asked you to come live in their society. They will sell you sugary products that you will like but be addicted to. These products, over time, will make you unhealthy and obese. You will feel depressed by how you look and feel. As you get sicker, you will pay more for medications that will do nothing to cure you. Does this sound like a place you want to live? We have news for you, you already live there.

## We are all addicts

With sugar being added to more and more products and people adding more sugar and sugary creamers to coffee, America is becoming addicted. If this proceeds, we predict that America will become more addicted to the wrong foods and as a consequence of this, it will get even sicker. We will need to build even more hospitals and there will be 3 or 4 drug stores in each community instead of the current 2.

Eleven foods with a high amount of hidden sugar are:[158]
1. **Fortune cookies**. Just one fortune cookie packs about 3.6 grams (5 grams are about 1 tsp) of added sugar.
2. **Flavored booze**. Exercise good judgment when you drink: One ounce of crème de menthe has 14 grams of added sugar; 53-proof coffee-flavored liqueur has 16 grams of added sugar per ounce (over 3 tsp).
3. **Baked beans**. A one-cup serving of canned baked beans with no salt added will cost you nearly 15 grams of added sugar.

4. **Dried, sweetened cranberries**. Without the sweetener, this fruit can be incredibly tart. But one serving-a third of a cup-of this treat will hit you with 25 grams of added sugar.
5. **Ketchup**. A favorite condiment, a single one-cup serving of regular-or low sodium-ketchup racks up nearly 40 grams of added sugar (WOW 8 tsp).
6. **Cream substitutes**. A one-cup serving of a liquid "light" cream substitute packs 22 grams of added sugar, while a one-cup serving of a powdered "light" cream substitute adds a whopping 69 grams.
7. **BBQ sauce**. A one-cup serving of this summertime favorite adds 9 grams of added sugar onto those ribs and chicken.
8. **"Reduced" salad dressings**. A one-cup serving of reduced-calorie French dressing heaps 58 grams of added sugar, and a one-cup serving of reduced-fat coleslaw dressing hits a home run with 103 grams of added sugar (reduce the fat and pile on the sugar).
9. **Lemonade**. A cup of lemonade powder has a massive 200 grams of added sugar. A single serving of the drink has almost 17 grams of added sugar.
10. **Flavored popcorn**. Think the added sweetener can't be that bad here? Fat-free-syrup caramel popcorn has 18 grams of added sugar per ounce serving.
11. **Granola bars**. Often deemed a healthful snack, some are tricky-a 1-ounce serving of a granola bar with oats, fruit, and nuts has 11 grams of added sugar.

How many of these high sugar death traps do you consume? How many are you feeding your family? **Q. Why? A. You are addicted to the sugar!**

## Breaking the Addiction

OK, we are addicted as a nation. So how do you start to reduce our sugar intake? Here are some tips.[159]
1. Try decreasing your intake of added sugar gradually. It can be difficult to suddenly cut all added sugar and refined carbohydrates. Try taking a week-by-week approach. One week, *add* less sugar to your morning coffee. Next week, replace your afternoon soda with bottled water. The following week, replace white bread with a whole grain alternative. Before long, you will find that the foods

(and drinks) you used to love now taste sickeningly sweet. And you will likely find it easier to keep your moods on an even keel, too.

2. Keep notes on your sugar intake in your journal or Day-Timer. How does decreasing your sugar levels impact your energy levels? Your mood during the day? Your ability to fall asleep at night? When do sugar cravings hit? It might be helpful to start with a Sugar Fast for a day or two. See how one day without added sugars affects you.
3. Make easy substitutions. Buy brown rice instead of white rice, for example. Brown rice has a nice, nutty flavor, and takes just a bit longer to cook. The next time you go to the store, experiment with all kinds of whole grain alternatives. You might find some new *family* favorites.
4. Keep healthy snacks readily available, and rely on a bit of protein in your snacks to keep your energy levels high. Keep a small bowl of nuts on the table, along with fresh fruits. When you are hungry for a mid afternoon snack, opt for lean protein and complex carbohydrates.
5. Indulge in moderation. If you are a chocoholic, treat yourself to a square of fine dark chocolate at the end of a long day. When the chocolate is quality, you won't feel the need to have more and you'll be more apt to take your time and savor it. When you do indulge in a sugary snack, keep it small, eat it slowly, and eat a bit of protein, too, to help moderate the blood sugar spikes and dips.
6. Dilute the fruit juice. If you or kids love fruit juice, try diluting it gradually to the point where you are just adding a splash to the top of water.
7. Become a sugar detective. You can start by knowing the alternate names for added sugars, often found in ingredients lists. These include any ingredient that ends in the suffix "-ose," including sucrose, dextrose, fructose, lactose, polydextrose, maltose, and galactose. Also, look for the following: corn syrup, high fructose corn syrup, honey, maple syrup, molasses, carob syrup, turbinado sugar, fruit juice concentrate, brown sugar, cane juice, cane sugar, evaporated cane juice, beet sugar, and sorbitol.
8. Avoid replacing added sugar with artificial sweeteners. Your best bet is to gradually reduce your taste for sweet foods, not to replace

them with chemical alternatives. On ingredients lists, look for sucralose, saccharin, aspartame, acesulfame K, and neotame.
9. Avoid the center of the supermarket as much as possible. That's where most of the processed foods are shelved. Instead, shop the perimeter for healthy, raw foods.
10. If you have young kids, go to the grocery store by yourself. You may be less apt to come home with sugary treats. Plus, you can take more time to examine the labels for hidden sugar. If your kids are grade school age or older, take them along and enlist their help as Sugar Detectives. Give them each a list of hidden sugars and artificial sweeteners and turn it into a game.
11. Carefully measure how much honey you put in your tea and how much sugar you put in your coffee. Aim to put in a bit less each day or each week until you are drinking it either unsweetened or with just a bit of sugar.
12. Buy items that are not sweetened, and *add* sugar only if you find that you need to. This will help you wean off the sugar gradually.
13. Steer clear of sugars for breakfast. When you start your day with a sugar blast and crash, you may find yourself in a vicious cycle for the remainder of the day. Start your day with healthy lean protein and complex carbohydrates. Try natural whole-grain breads and cereals for breakfast, along with a low fat protein, such as skim milk, cottage cheese, or yogurt.

**"NOTE: We do not believe everything in this quote; even small amounts of milk and cheese are unhealthy. All cereals except whole grain oatmeal turn to sugar quickly in your system."**

## Breaking the Sugar Hold

If you remove sugar from your diet, you will find that in about 3-4 days it has less of a hold on you. If you go back, the addiction starts all over again. There are a lot of people that look at Alcoholism as a bad and embarrassing disease. It is embarrassing to most, but *"Sugarism"* is much worse. It kills more Americans and cripples many more through diabetes. No one looks at sugar as embarrassing. It is the *sweet* disease. What do you offer your grandkids? Is it sweets or vegetables if they have been good? We are taught from childhood that sweets are a reward.

I have seen my own family treat sugar as some great reward. How many times have you seen someone say to a child "Eat your vegetables and you can have a cookie (or candy or cake or ice cream)?" What these people are really saying is eat that horrible tasting vegetable and I will give you some poison that tastes so good. It reminds us of the Hansel & Gretel story where the wicked witch lures them into her lair. Sugar is poison to our bodies and health. Yes there are natural sugars, but they enter the body much slower (see Chapter 8: Glycemic Index) because of their packaging. A small amount of sugar may not be harmful to your health but on average we are consuming 32 teaspoons of sugar a day. That is pretty sweet.

## *Science*

Princeton University researchers say they have the proof of sugar addiction in animal studies.[160] Other scientists suggest our overeating may cause similar neurological changes as does drugs.[161] Other studies suggest most Americans are addicted to sugar but when asked will insist, "I am NOT addicted, I can stop at any time". This sounds a lot like alcoholics, drug users and smokers. The test is to try to stop all sugar intake for just a week or two. It's a lot harder than it sounds. There are many studies that say sugar is not addictive. Remember the sugar industry is very strong and pays scientists to do research that disprove any addiction. Does this sound familiar? Remember the cigarette companies said nicotine is not addictive?

## *Conclusion*

We are fed sugars from infancy as a reward. Our mothers and grandmothers are the guilty parties. How can we resist something that is fed to us all our lives as a reward? Raw sugar is a relatively new substance to human diet. The ramifications of eating too much of it are everywhere. They include:
- Obesity
- Diabetes
- Cancer
- Depression

No one really wants to hurt their kids or grandkids but we must take a strong look at our reward systems and diet. If we don't, they are condemned to a life of suffering and possibly illness.

# Chapter 14:

# How do you get your Protein

## *Introduction*

In Chapter 7: Vegetarianism, we discussed that protein is in every living thing. Amino acids make up protein and they are the basis of all life on earth and probably everything living in the Universe. Proteins are so important to life. If we think of all life as being made up of cells (except viruses which are not cellular but do contain DNA or RNA and Protein); and we know a cell is constructed from protein. We make all the amino acids we need except for nine called the essential nine amino acids (one of the nine is not needed after early childhood; so there are really only eight essential amino acids in adult humans).

## *Basics*

**Prions can cause Disease**

Prions are proteins that have taken an unusual shape. Proteins are a chain of amino acids but they are also folded into shapes. How they fold matters. These deformed prions can be dangerous to health. Unlike viruses and bacteria that have DNA or RNA, prions are only a protein module.[162] The types of proteins we ingest do matter. Microwaves can change the form of protein!

Russia banned microwaves because proteins can be dangerously changed, causing destruction of nutrients and increasing risk of cancer.[163] Another study showed that microwaving vegetables destroys up to 97%

of the nutritional content (vitamins and other plant-based nutrients that prevent disease, boost immune function and enhance health).[164]

## Does Protein in Food Trigger Epigenetics?

We eat food made of proteins. Proteins are chains of amino acids folded in certain ways. They can get folded wrong. Our digestive system breaks protein down into amino acids and chains of amino acids for reuse in the cell. We also know that amino acids, chains of amino acids, proteins and prions can all cause epigenetic changes in our cells. These changes can block the activation of DNA genes we need or they can activate other dominant genes we don't need. Either way it affects out health in very complex ways. **We really are what we eat!**

It would seem that proteins and the amino acids they contain do have an effect on health. They act as triggers to turn on or off certain genes. The combination of enabled and disabled genes determines our health. The new epigenetic genome project will try to study these effects, but it will be a daunting task for sure. We have already read that many scientists are saying the epigenetic genome will provide a new series of drugs to fight disease. Maybe some smart scientists will realize we should be looking for cures, not more pills, to cover up or camouflage the symptoms. Epigenetics is an important study of how the food and environment around us affects our health and well being. Let's really improve the quality of life, not add more drugs.

### *Ingested Protein*

The food we eat gets broken down into its amino acids and chains of amino acids. Some may remain as long chains (proteins). Remember proteins and amino acid chains can act as messengers causing an epigenetic reaction in our cells. We believe that eating foods from sources close to us on the evolutionary scale has a higher probability of turning *ON* or *OFF* genes that normally wouldn't have been. Animals (meat) are clearly close to us evolutionary wise 90%+ the same DNA. Plants are farther away (50%+ the same DNA). If this is true, it explains how certain foods, such as meats, can adversely affect us. Early man didn't walk around and worry about proteins and excess amounts. Nature grew us from what we were eating over millions of years. Our DNA and epigenetics were built based upon our diets. These diets did not change much over this time. Suddenly after

World War II, we drastically changed everything including our health. Now we spend time and energy building more hospitals to handle the growing numbers of sick people.

## *Advanced*

### The Distance down the Evolutionary Chain

The type of proteins we ingest would seem to matter. It may be that we were designed to eat proteins farther down the evolutionary chain than us. Fish is farther down than meats but plant based life is even farther away. We believe that eating a variety of plant based foods is essential to good health. We see studies that show a certain type of food has a positive effect on certain diseases or health conditions. However, to get this positive effect, we don't have to study these and be aware of what does what. Just eat a variety and know you are obtaining all the nine essential Amino Acids as well as a variety of antioxidants. It shouldn't be a chore but a pleasure to consume healthy foods.

### Fruits & Vegetables

The most powerful food you can eat is plant based foods. Even if you choose to eat a small amount of meat in your diet, you should be careful to eat plenty of plant based foods such as fruits, vegetables and nuts. These foods have evolved with us for millions of years. Our bodies have developed mechanisms to fight bacteria, viruses and diseases by using the proteins in plant based foods. These proteins consist of the nine essential amino acids we need to live. Never underestimate the power of plants on our lives. If you are a vegetarian or even a vegan you still need to be careful of not eating much processed foods or junk foods. Vegetarians have about half the cancer risk of meat-eaters.[165] How important is not getting cancer to you?

### Meats

The dangers of red meat have been made public for a while now but all meat may trigger a bad epigenetic reaction. Consumption of meat is linked to cancer, heart disease, diabetes and obesity. The protein in the meat we eat is very similar to the protein in humans. This protein may be a trigger for bad genetic reactions that can cause disease. It is a very complex area and the interactions are different in each of us. In the book, the China Study, T. Colin Campbell [166] says he found that consuming more than 10% animal protein triggered cancer and heart disease. Animal protein is not just meat but eggs, milk, yogurt and butter as well. How much animal protein do you ingest daily? Count your total calories and those calories from all animal products daily. Then compute the percentage by dividing the animal protein calories by the total calories. If you are eating more than 10% you may be adversely affecting your health.

## Fish

Many people consider fish to be animal protein as well. It may have negative affects as well on our genetics. Fish can also have high levels of mercury. In our belief, fish is farther down the evolutionary scale from us than meat. Its protein is less likely to interfere negatively with our epigenetics.

## Milk

The Irish Medical Times did a study of the diagnosis and management of cow's milk protein allergy.[167] They found that the proteins in milk can affect our bodies and causes us to develop allergies. Clearly milk protein is a likely epigenetic trigger in humans, creating negative effects. After all it was designed to trigger epigenetics in cows not humans!

Some studies have linked milk to breast cancer in woman and to Type 1 diabetes[168]. Cow's milk contains insulin that could trigger an autoimmune reaction in a human causing Type 1 Diabetes. Some woman have opted to have their daughter's breasts removed to save them from what they believe is a family genetic disposition to breast cancer. Stopping epigenetic triggers like milk is a lot more humane and easier on the child.[169] Remember we are NOT predisposed by our DNA to any condition. It is a combination of our DNA and the epigenetics that trigger our genetic genes. These triggers, unlike DNA, can be changed in our lives.

## *Science*

Protein is defined as: Any of a group of complex organic macromolecules that contain carbon, hydrogen, oxygen, nitrogen, and usually sulfur and are composed of one or more chains of amino acids. Proteins are fundamental components of all living cells and include many substances, such as enzymes, hormones and antibodies that are necessary for the proper functioning of an organism.[170] The key here is protein *is* what makes up the cell walls of all living organisms on Earth! This includes plants (vegetables & fruit) and of course meats.

## *Conclusion*

We must be aware of the epigenetic triggers that can affect us. Science is not yet at the point of mapping these triggers and understanding how they work but they know the triggers exist. Proteins are critical to our health. We are made of proteins. They are triggers that can cause positive or negative epigenetic reactions. Proteins come from ALL foods (plant and animal based). Animals get their protein from plants. When we eat meat, we are getting proteins from plant based sources that the animal ate or another animal eaten that ate plant based foods.

We believe foods closer to us on the evolutionary chain can cause more harm than foods farther down the chain. Plant-based foods are much farther away from us (evolutionarily) than animal-based foods are.

Vegetarian diets are extremely healthy and provide all the nutrients and antioxidants nature made for us.

# Chapter 15:

# Diseases and Cures

## *Introduction*

"The poorest man would not part with health for money, but the richest would gladly part with all their money for health."—*Charles Caleb Colton*

"The first wealth is health."—*Ralph Waldo Emerson*

"Take care of your body. It's the only place you have to live."—*Jim Rohn*

"Time and health are two precious assets that we don't recognize and appreciate until they have been depleted."—*Denis Waitley*

Most of us don't think about death until we get older and closer to it. We also seldom think about illness while we are healthy but when we are diagnosed with an illness, we can't stop thinking about it. Making changes now while you are healthy can ensure you will remain healthy your entire life. A real world example is your automobile. If you drive it using your brakes a lot, they can be prematurely worn down. They may fail in a situation where you need them at full capacity to stop quickly. If we misuse our bodies they may also fail us prematurely. Why not take control of the things we can control?

## Basics

We have been talking about how bad foods and bad environment can cause diseases. Now we will look at the various diseases that afflict so many of us. Here are some devastating statistics from Disabled World (originally from Center for Disease Control report on state of aging:[171])

## CDC Report on State of Aging

- 88% of those over 65 years of age have at least one chronic health condition
- Nearly 40% of deaths in America can be attributed to smoking, physical inactivity, poor diet, or alcohol misuse
- Almost 20% of older Americans suffer from a mental disorder that is NOT part of normal aging
- Even though cognitive decline (becoming senile) is NOT a normal part of aging, many people assume it is and 25% of the elderly have experienced it.
- Arthritis and related conditions are the leading cause of disability in the U.S.
- Heart disease is still the leading cause of death with cancer coming in second
- 20% of people over 65 have diabetes
- Falls are the leading cause of injuries—one in three people over 65 fall each year

And guess what? All of these are preventable! I guess that's the good news but it sounds like a wake-up call to us. Statistics show that in the over 65 age group, 30% have 3 or more chronic diseases. These multiple diseases can complicate diagnosing as one could mask symptoms of another, especially when adding the complication of taking multiple drugs with their various side effects.

## How to Prevent Aging Diseases

A change in lifestyle and behaviors is REQUIRED if we are going to live healthy lives as we age. Old age is NOT an option but illness is!

## Quit ALL Tobacco Use

This should be a no brainer. We've been hearing for years how smoking affects our health and the health of those around us that don't smoke. Yet there are still many that think the occasional cigar or chewing tobacco doesn't count.

A research team from Australia and San Antonio, Texas, analyzed white blood cell samples of 1,240 people, ages 16-94, who were participating in the San Antonio Family Heart Study. They found that the self-identified smokers in the group—297 people—were more likely to have unusual patterns of "gene expression (epigenetics)" related to tumor development, inflammation, virus elimination, cell death and more. A gene is expressed when it codes for a protein that then instructs, or kick-starts, a process in the body. The authors of the study found cigarette smoke could increase or decrease the level of expression of 323 genes.[172] A bad habit such as smoking can cause a drastic epigenetic change in our bodies resulting in illness and death.

## Regular Physical Activity

You don't have to become Jack LaLane—even regular walking will do you good. Strength training of some type will help in a variety of ways. By age 75, about one in three men and one in two women do not engage in any physical activity. That's really sad.

## Good Nutrition

This is simple basic stuff. Low fat, lots of fruits and vegetables. Sadly, less than one-third of adults over 65 meet the minimum of 5 servings a day recommendation. Plants evolved with us and give us the missing nine amino acids as well as plenty of antioxidants which we need.

## Increase Your Social and Mental Activity

Research has shown that the more socially active older person has less chance of developing depression and other mental disorders associated with aging. Do something as simple as crossword puzzles will help to maintain your mental acuity. Our social interactions with each other, seems to also have an epigenetic effect on our cells.

Of course, there's a lot more you can do to live a healthy, happy life as you age. There are some surprising good news and studies being done in the field of anti aging. Keep yourself informed.

## Cost of Chronic Illness

Epigenetics does affect our health. Some affects we can do nothing about but food is something we can change and learn to enjoy. The American economy is being crippled by health care costs. We can't continue on this path without totally breaking the system.[173]

Christine Houghton, PhD, Nutritional Biochemistry & clinician, says we are in a paradigm shift empowering individuals to take back responsibility for their own health! Foods such as broccoli has a substance called Sulforaphane that has been shown to affect at least 200 epigenetic changes. These changes improve our resistance to diseases and help cell defenses.[174]

In the U.S., chronic illness has risen to epidemic proportions over the last 50-60 years. The incidences of cancer, heart disease, diabetes, obesity, arthritis, autoimmune issues, depression and anxiety disorders, chronic fatigue, chronic pain, infertility, insomnia and other sleep disorders, acid reflux, constipation, decreased sex drive, Alzheimer's, dementia, ADD/ADHD and many other conditions continue to sky rocket, despite our "best" efforts.

It has been stated by leading health experts, as well as economists, that the problem of chronic illness, if it continues to grow at its current rate, will cripple the U.S. economy and health care system in just a few short years. We can't support the majority of our citizens being ill and unable to fully contribute to the health and energy of the country.

Recent reports reveal that 80% of all doctor's and clinical visits, all prescription drugs, and all expenditure in health care across the board is the result of chronic illness. When we look at all age groups in our population, including the youngest children, 46-52% of the entire population is currently diagnosed with a chronic illness. That's insane.

If our cells are little factories doing things dictated by our genes then they need certain raw materials to function properly. It may be we all have genes to fight diseases but they need certain inputs to work. Food is one critical input needed. As we eat certain foods like broccoli we cause certain epigenetic changes to happen. These changes help us battle toxins and viruses in our bodies. Without the broccoli we might not be able to start these gene defenses. Now imagine thousands maybe hundreds of thousands of epigenetic changes going on in a person that eats healthy as compared to a person that does not. The person ignoring healthy food is doomed to pay the consequences of not providing their cells with what they need to remain healthy.

## Healthy Eating vs. Drugs

The medical industry at large has us believing that taking drugs and chemotherapy will cure our cancer. In some cases it does temporarily and in others it does not. Chemotherapy is poison! We are killing the cancer cells and ourselves with the hope we will kill the cancer before it kills us. We have said it before and it's worth saying again: Eating healthy is not bad for you. No one will say eating more fruits and vegetables are bad for you. Some people have said to me that eating healthy and organic is too expensive. They cannot afford *it*. We always tell them you have a choice. **Pay now with healthy foods or later with drugs and medical procedures. The choice is yours.**

## Cancer

The words *"you have cancer"* are one of our greatest fears. We all know family and friends that have it and have died of it. It is almost synonymous with old age in this country. Cancer is not a simple disease to understand. Some say it is many diseases mixed together. The genes that cause cancer may actually be useful genes. The fertilized egg must grow very rapidly early on. This is like a cancer. It then turns off and growth happens at a more controlled rate. Do we all have cancer genes? The answer is probably yes! The real issue is not if we have the gene but if we have turned it on or have turned off some other gene that controls cancer. The answer, most likely, is it is a very complex set of gene interactions. Some being turned on and others being turned off that cause most types of cancer.

According to the American Cancer Society over 1 million people currently get cancer each year. One out of every two men and one out of every three women suffer from this disease. Why does one person get cancer and another does not is a complex question. The answer lies in your genetic make-up (both genes & epigenetics), environmental factors, stress, or any number of other factors.[175] Our genetic make-up is both our genes (which are very similar to every other human on Earth) and our epigenetics. Read everything you can on epigenetics, healthy eating and thinking. These factors will save your life and even prevent cancer from turning you into a statistic.

We know by now that epigenetics causes genes to turn on and off. We also know that our food, thoughts and beliefs control epigenetics. Eating the right food with plenty of natural fruits and vegetables (not processed) will increase the probability of not activating cancer. How much easier can it be to be healthy? We are being marketed into an unhealthy lifestyle because it makes big business a lot of money. Answer one simple question honestly. Do you want to be healthy? If the answer is yes, read on.

*How do genes affect cancer growth?*

The discovery of certain types of genes that contribute to cancer has been an extremely important development for cancer research. Over 90 percent of cancers are observed to have some type of genetic alteration. A small percentage (5 percent to 10 percent) of these alterations are inherited, while the rest are sporadic, which means they occur by chance or occur from environmental exposures (usually over many years). There are three main types of genes that can affect cell growth, and are altered (mutated) in certain types of cancers, including the following:[176]

- **Oncogenes**
  These genes regulate the normal growth of cells. Scientists commonly describe oncogenes as similar to a cancer "switch" that most people have in their bodies. What "flips the switch" to make these oncogenes suddenly become unable to control the normal growth of cells and allowing abnormal cancer cells to begin to grow, is unknown. It is an epigenetic switch so it probably is diet, stress, environment or a combination of them.
- **Tumor suppressor genes**
  These genes are able to recognize abnormal growth and reproduction of damaged cells, or cancer cells, and can interrupt their reproduction until the defect is corrected. If the tumor suppressor genes are mutated, however, and they do not function properly, tumor growth may occur.
- **Mismatch-repair genes**
  These genes help recognize errors when DNA is copied to make a new cell. If the DNA does not "match" perfectly, these genes repair

the mismatch and correct the error. If these genes are not working properly; however, errors in DNA can be transmitted to new cells, causing them to be damaged.

Usually the number of cells in any of our body tissues is tightly controlled so that new cells are made for normal growth and development, as well as to replace dying cells. Ultimately, cancer is a loss of this balance due to genetic alterations that "tip the balance" in favor of excessive cell growth.

## *Cancer Treatments*

Currently cancer can be treated in many ways: radiation, chemotherapy, or surgery. Wouldn't it be better to not get cancer in the first place? Then you wouldn't need any of the above three drastic measures, better known as torture, to save your life. Cancer IS NOT inevitable. Don't let anyone tell you it is part of living longer or you have to die someway. Live healthy and die healthy; that's what is natural. When we have radiation, chemotherapy or surgery it targets a part of cancer in our body but does not cure it. If our lifestyle still enables bad genes through epigenetics, we will get the cancer back. Typically these treatments don't remove all of the cancer.

## *Heart Disease*

Each year, heart disease is at the top of the list of the country's most serious health problems. In fact, statistics show that cardiovascular disease is America's leading health problem, and the leading cause of death. Consider the most recent statistics released by the American Heart Association (AHA):[177]

- At least 71 million people in this country suffer from some form of heart disease.
- One person in three suffers from some form of cardiovascular disease. This includes high blood pressure—65 million; coronary heart disease—13 million; stroke—5.5 million; congenital cardiovascular defects—1 million; and congestive heart failure—5 million.
- Rheumatic heart disease / rheumatic fever kills 3,554 Americans each year.
- Almost one out of every 2.7 deaths results from cardiovascular disease.
- More than 2,600 Americans die of cardiovascular disease each day, an average of one death every 34 seconds.

- Cardiovascular disease is the cause of more deaths than the next five causes of death combined, which are cancer, chronic lower respiratory diseases, accidents, diabetes mellitus, and flu/pneumonia.
- It is a myth that heart disease is a man's disease. In fact, cardiovascular diseases are the number one killer of women (and men). These diseases currently claim the lives of nearly a half a million females every year.
- About one-third of cardiovascular disease deaths occurred prematurely (before age 75, the approximate average life expectancy in the year the study was done).
- On average, someone in the US suffers a stroke every 45 seconds; someone dies every 3 minutes from stroke.
- Stroke is a leading cause of serious, long-term disability that accounts for more than half of all patients hospitalized for a neurological disease. Stroke deaths have been increasing in recent years.

## *Diabetes*

What is Diabetes?

Diabetes is a metabolic disorder characterized by a failure to secrete enough insulin, or, in some cases, the cells do not respond appropriately to the insulin that is produced. Because insulin is needed by the body to convert glucose into energy, these failures result in abnormally high levels of glucose accumulating in the blood. Diabetes may be a result of other conditions such as genetic syndromes, chemicals, drugs, malnutrition, infections, viruses, or other illnesses.[178]

The three main types of diabetes—type 1, type 2, and gestational—are all defined as metabolic disorders that affect the way the body metabolizes, or uses, digested food to make glucose, the main source of fuel for the body.

What is Prediabetes?

In prediabetes, blood glucose levels are higher than normal but not high enough to be defined as diabetes. However, many people with prediabetes develop type 2 diabetes within 10 years, states the National Institute of Diabetes and Digestive and Kidney Diseases. Prediabetes also increases the risk of heart disease and stroke. With modest weight loss and moderate physical activity, people with prediabetes can delay or prevent type 2 diabetes.

How does Diabetes affect Blood Glucose?

For glucose to be able to move into the cells of the body, the hormone insulin must be present. Insulin is produced primarily in the pancreas, and, normally, is readily available to move glucose into the cells.

However, in persons with diabetes, either the pancreas produces too little or no insulin or the cells do not respond to the insulin that is produced. This cause a build-up of glucose in the blood, which passes into the urine where it is eventually eliminated, leaves the body without its main source of fuel.

How do the three main Types of Diabetes Differ?

Although the three main types of diabetes are similar in the build-up of blood glucose due to problems with insulin, there are differences in cause and treatment:

- **Type 1 diabetes**
  Type 1 diabetes is an autoimmune disease in which the body's immune system destroys the cells in the pancreas that produce insulin, resulting in no or a low amount of insulin. People with type 1 diabetes must take insulin daily in order to live.
- **Type 2 diabetes**
  Type 2 diabetes is a result of the body's inability to make enough, or to properly use, insulin. Type 2 diabetes may be controlled with diet, exercise, and weight loss, or may require oral medications and/or insulin injections.
- **Gestational diabetes**
  Gestational diabetes occurs in pregnant women who have not had diagnosed diabetes in the past. It results in the inability to use the insulin that is present and usually disappears after delivery. Gestational diabetes may be controlled with diet, exercise, and attention to weight gain. Women with gestational diabetes may be at higher risk for type 2 diabetes later in life.

Complications of Diabetes:

Diabetes is the sixth leading cause of death among Americans, and the fifth leading cause of death from disease. Although it is believed that diabetes is under-reported as a condition leading to or causing death, each year, more than 200,000 deaths are reported as being caused by diabetes or its complications. Complications of diabetes include eye problems and blindness, heart disease, stroke, neurological problems, amputation, and impotence.

Because diabetes (with the exception of gestational diabetes) is a chronic, incurable disease that affects nearly every part of the body, contributes to other serious diseases, and can be life threatening, it must be managed under the care of a physician throughout a person's life.

If we tried to sell you a drug that would make you sick, impotent, go blind and run the risk of losing your limbs; would you buy it and take it? But most Americans do buy and take it. The drug is called sugar!

**Preventing Diseases**

*Heart Disease*

The food guide pyramid is a guideline to help you eat a healthy diet. The food guide pyramid can help you eat a variety of foods while encouraging the right amount of calories and fat. The United States Department of Agriculture (USDA) and the US Department of Health and Human Services have prepared the following food pyramid to guide you in selecting foods.[179]

**MyPyramid.gov**
STEPS TO A HEALTHIER YOU
*Source: US Department of Agriculture*

The Food Pyramid is divided into 6 colored bonds representing the 5 food groups plus oils:
- **Orange** represents grains: Make half the grains consumed each day whole grains. Whole-grain foods include oatmeal, whole-wheat flour, whole cornmeal, brown rice, and whole-wheat bread. Check the food label on processed foods—the words "whole" or "whole grain" should be listed before the specific grain in the product.
- **Green** represents vegetables: Vary your vegetables. Choose a variety of vegetables, including dark green—and orange-colored kinds, legumes (peas and beans), starchy vegetables, and other vegetables.
- **Red** represents fruits: Focus on fruits. Any fruit or 100 percent fruit juice counts as part of the fruit group. Fruits may be fresh, canned, frozen, or dried, and may be whole, cut-up, or pureed. Fresh whole fruits are best for overall general health.
- **Yellow** represents oils: Know the limits on fats, sugars, and salt (sodium). Make most of your fat sources from fish, nuts, and vegetable oils. Limit solid fats like butter, stick margarine, shortening, and lard, as well as foods that contain these.
- **Blue** represents milk: Get your calcium-rich foods. Milk and milk products contain calcium and vitamin D, both important ingredients in building and maintaining bone tissue. Try converting to Soy Milk instead of Cow's milk.
- **Purple** represents meat and beans: Go lean on protein. Choose low fat or lean meats and poultry. Vary your protein routine—choose more fish, nuts, seeds, peas, and beans.

Activity is also represented on the pyramid by the steps and the person climbing them, as a reminder of the importance of daily physical activity.

## *Cancer*

Here are some tips by the Mayo Clinic staff about cancer prevention:[180]

You've probably heard conflicting reports in the news about what can or can't help you in terms of cancer prevention. The issue of cancer prevention gets confusing—sometimes what's recommended in one report is advised against in another. What you can be sure of when it comes to cancer prevention is that making small changes to your everyday life

might help reduce your chances of getting cancer. Try these seven cancer prevention steps.
- **Cancer prevention step 1: Don't use tobacco**
- **Cancer prevention step 2: Eat a variety of healthy foods**

The American Cancer Society recommends that you:
- Eat an abundance of foods from **plant-based sources**. Eat five or more servings of fruits and vegetables each day. In addition, eat other foods from plant sources, such as whole grains and beans, several times a day. Replacing high-calorie foods in your diet with fruits and vegetables may help you lose weight or maintain your weight. A diet high in fruits and vegetables has been linked to a reduced risk of cancers of the colon, esophagus, lung and stomach. Why do we wait until we have cancer to eat this way?
- Limit fat. Eat lighter and leaner by choosing fewer high-fat foods, particularly those from animal sources. High-fat diets tend to be higher in calories and may increase the risk of overweight or obesity, which can, in turn, increase cancer risk.
- Drink alcohol in moderation, if at all. Your risk of cancers, including mouth, throat, esophagus, kidney, liver and breast cancers, increases with the amount of alcohol you drink and the length of time you've been drinking regularly. Even a moderate amount of drinking—two drinks a day if you're a man or one drink a day if you're a woman and one drink a day regardless of your sex if you're over 65—may increase your risk.
- **Cancer prevention step 3: Stay active and maintain a healthy weight**
- **Cancer prevention step 4: Protect yourself from the sun**
- **Cancer prevention step 5: Get immunized**
- **Cancer prevention step 6: Avoid risky behaviors**
- **Cancer prevention step 7: Get screened**

Methyl groups in foods like beets, green leafy vegetables and legumes, when eaten in large quantities have shown a positive effect on cancer. [181] Remember Methyl groups are the triggers of epigenetics that enable or disable genes. The evidence is abundant; eat your plant based foods often!

## *Diabetes*

Here are some tips by the Mayo Clinic staff about diabetes prevention:[182]

Tweaking your lifestyle could be a big step toward diabetes prevention—and it's never too late to start. Consider these 5 tips:
- **Tip 1: Get more physical activity**
- **Tip 2: Get plenty fiber**
  It's rough, it's tough—and it may reduce the risk of diabetes by improving your blood sugar control. Fiber intake is also associated with a lower risk of heart disease. It may even promote weight loss by helping you feel full. Foods high in fiber include fruits, vegetables, beans, whole grains, nuts and seeds.
- **Tip 3: Go for whole grains**
- **Tip 4: Lose extra weight**
- **Tip 5: Skip fad diets and make healthier choices**
  Low-carbohydrate, low-glycemic load or other fad diets may help you lose weight at first, but their effectiveness at preventing diabetes isn't known; nor is their long-term effects. And by excluding or strictly limiting a particular food group, you may be giving up essential nutrients. Instead, think variety and portion control as part of an overall healthy-eating plan.

## *Advanced & Science*

### Body Signs

Our body gives us notification of bad things happening to it. We must learn to listen to our bodies better.
- **Fever**—Most people get so worried about a fever and take medication for it. Although very high fevers can be very dangerous, a normal fervor is our body's way of fighting infections, bacteria and viruses. It raises our temperature to kill off the invading enemies. By taking medicine to lower the temperature back down, we actually fight our body's defense mechanisms and allow the invaders easier access. It's sometimes better left untreated. Fever seems to play a key role in helping your body fight off a number of infections.[183]
- **Pains**—Can sometimes be our body's way of telling us we are doing something wrong. Maybe we are over stressing our muscles.

Pain, though disruptive, is a warning signal that the body needs attention. Sometimes the answer is simple, such as rest for muscle overuse. Yet, if generalized pain has you down, it is possible to get rid of body aches with home therapy and/or medical support, depending on its cause. [184]
- **Disease**—Is often our body's way of saying "I can't fight it anymore" without proper nourishment or from too much stress. When we mistreat our bodies, just like a car, they often fail us. It is like a civil war inside our bodies.[185] Remember we are made up of a vast city of cells all doing their jobs for our (and their) survival. They need some things from the world like the nine essential amino acids, anti-oxidants and trace minerals (from our food). They need to have an environment that is not always in stress. They need our minds working with them not against them.

## Statistics

The top ten men's and women's diseases are listed in Appendix E: Top Ten Diseases. These two lists are from the Mayo Clinic staff.

## *Conclusion*

> *"Most diseases are the result of medication which has been prescribed to relieve and take away a beneficent and warning symptom on the part of Nature."*—Elbert Hubbard

Diseases are complex and their causes are a combination of many factors. Those factors are the things that affect our epigenetics. We have stated this many times in this book. It is critical to understand that our health is a function of:
- Food
- Environment
- Mood
- Stress
- Mental thinking
- Beliefs

Some of these come from things we cannot control like radiation, second hand smoke, etc but most come from things well under our control.

We are healthy because we want to be healthy and follow a healthy way of life. Conversely we are unhealthy because we want to be and don't follow a healthy way of life.

Most of us are concerned about diabetes, high blood pressure and cholesterol. We get them checked religiously and most people over a certain age are on medications for one or more of these. Dr. Dean Ornish, M.D. in this book "The Spectrum" covers these topics and shows how they can be prevented and reversed by changes to diet and lifestyle. Here is what he says about each one:

- **Diabetes**—He states that diabetes is totally preventable and if you have it you can reverse it. He provides a diet that moves you from a high fat meat diet to a vegetarian one.[186]
- **High Blood Pressure**—diet, exercise and meditation has lowered blood pressure in many people. He says that eating more vegetables causes our bodies to produce Nitric Oxide which in turn relaxes our arteries and lowers our blood pressure. Dark Chocolate does the same through flavonols which are converted to nitric oxide. [187]
- **Cholesterol**—Dietary changes have a direct affect on your cholesterol. [188] The amount of exercise (not the intensity) seems to have a greater effect on harmful cholesterol levels. [189]
- **Weight**—Diet obviously affects our weight but so does stress. Dr Ornish says "chronic stress causes the body to secrete a cascade of hormones from the hypothalamus to your pituitary gland in your brain to other organs in your body (such as your adrenal glands and your thyroid), which, in turn, secrete hormones such as glucocorticoids and insulin, which cause you to gain weight and accumulate fat tissue, especially around your belly, where it's most harmful and least attractive."[190]

## Disease Solutions

Does anyone believe we are really working on a solution to cancer or AIDS or Diabetes? No way! Can you imagine the revenue loss in drugs if they suddenly cured it? The drug companies would go out of business. Their goal is to make money not lose it. The weight management companies are the same. Do you really believe any of them wants you to lose weight and keep it off? To do so would put them out of business. Their goal is also to make money. You have to follow the dollars and understand that all businesses are there to make money, not lose it.

We are living in a time where we have significantly changed our food supply, our environment and the stresses on us. We are marketed to buy the wrong things for a healthy lifestyle. We even make up names for it like *comfort foods*. We get sicker and our quality of life goes down. We are then sold drugs that do not cure the diseases but may slightly extend our life. So we live longer in a low quality of life causing more stress. It's a viscous loop. Who is the real victim here? We are!

# Chapter 16:

# Epigenetics—Stress

## *Introduction*

Stress and health are much related. Our modern society places a lot of stress on us. Our ancient ancestors had stress when attacked by a predator that ended either because they were eaten or they got away from the predator and the stress ended. Their stress came in short bursts. Stress gave them the power to try to overcome their plights. Today our stress is always there. We are stressed because we woke up late, had nothing around for breakfast, got into traffic, had our boss yell at us, had to deal with others, forgot a birthday or anniversary, etc. It never ends. This was to help our old *Fight vs. Flight* paradigm. Stress was designed to focus us on one thing so we could win and survive another day. Nature never intended it to be a constant state of mind. We must learn to not allow stress to take control of us!

## *Basics*

### Lowering Our Stress

The Center for Young Woman's health says we can lower our stress by:[191]

1. **Simplify.** You may feel like you're not in control of everything that's expected of you. It's up to *you* to decide what you can do, and what you can't. To help simplify your life, sit down and make a list of everything you feel you need to do. Then separate all the items on the list into these three sections:

a. These can wait
   b. These are pressing
   c. Do these TODAY

   If you see that there's just too much to do TODAY, you'll have to cut down on some activities to make your schedule more manageable.

2. **Exercise** is a great way to lower your stress. While exercising, you can focus on what you're doing with your body, which helps free your mind from other worries. Vigorous exercise also triggers the release of chemicals in your body called endorphins, which make you feel happier and more relaxed. You don't have to be a super-athlete to exercise. Even something as basic as walking for half an hour can help you relax and improve your mood. Or, you can sign up for a class at your local YWCA or YMCA. Choose something fun and friendly, such as dancing, volleyball, or swimming.

3. **Yoga, Tai Chi, & Qigong.** These types of movement from India and China use stretch and poses for flexibility, strength, concentration, and relaxation. Yoga emphasizes flexibility and strength, while Tai Chi and Qigong help with concentration, balance, and patience. You can do any of these exercises in a class at your local YWCA, YMCA, dance center, or at home on a towel or mat. If you're shy about taking a class, you can check a DVD out of the library, or find one on TV and try the movements at home.

4. **Take a Break.** Sometimes your tired brain is just craving a little time off from your busy day. Stop what you're doing, and find a quiet spot where you can put your feet up. Drink some tea (without caffeine), or take a bath. Read a book or magazine, or even watch TV, if it's a non-stressful show. These things sounds so basic, you might think, "why bother?", but when your feet are up, your stress level drops.

5. **Meditation and Prayer** offer you ways to calm, focus your thoughts, and feel more positive. There are many styles of meditation which have grown out of spiritual practices around the world. Meditation involves sitting still in a quiet place, focusing your thoughts on your breath or on a slow chant, and trying to be aware of what is

going on in the present moment, instead of stressing about the past or freaking out about the future. With prayer you focus on feeling connected to a higher spiritual power, and on wishes and hopes you may have for yourself or people you care about. Get in touch with your local church, temple, Yoga center, or Buddhist center about a prayer or meditation group. If you're shy about attending a group, you can check DVD's out from the library about different meditation and prayer techniques.

6. **Massage** can work wonders on a stressed-out body. A gentle massage can untie knotted muscles, and make you feel relaxed all over. A professional massage can be expensive, but even a simple foot-rub or shoulder-rub from a good friend can take the edge off your stress.

7. **Journaling.** If you enjoy writing, this can be a good way to de-stress. Write down what's been happening with you on a daily basis. If you're facing a scary situation, imagine the best-case and worst-case scenarios. Write about the worst thing that could happen if everything goes wrong. Then write about the wonderful things that would happen if everything goes right. By letting your mind explore all the possibilities you'll feel less stressed. Another thing you can do in your journal is write a letter to someone you're really mad at. Later on you can edit it and actually mail it, but sometimes it helps just to write it down.

8. **Have a good cry.** You may know that little kids get upset easily, cry and make a fuss, and then get over it pretty quickly. This approach can work for you too. At the end of a particularly hard day, if you find yourself crying to a supportive friend, family member, or to your pillow, this can help you de-stress. In our culture we often try to convince people not to cry, as if it were a sign of weakness, but that's really not true. If crying helps you communicate your frustration, vent your stress, and get some support, than there's nothing wrong with a good cry every now and then.

9. **Sleep.** Teens in our culture are notoriously sleep-deprived on a daily basis, and even just a few nights in a row of not-enough-sleep can make you feel irritable and nervous. You actually need *more*

sleep at this time in your life (about 9 hours per night) than you will as an adult. Although your school schedule and social life make it difficult, try to put sleep at the top of your priority list, right up there with eating healthy foods and watching your favorite TV shows. If you can squeeze in an additional hour or two of sleep per night, you'll feel a lot better, and your overall stress level will drop.

## *Advanced*

### **Support Groups**

Support groups are groups of people sharing the same problem. These people join these groups to share life and its challenges. We also learn from each other and become stronger because of what we learned. Maybe it's the loss of a loved one. Everyone has been through it and can offer how they got through it. It can help people who are *stuck* in grief. Check your town or church or synagogue to see if there are life groups that might help you. We are created to be with one another in life's journey. Life Groups allows this to happen. Share your grief with others. It is a way of sharing it but also getting rid of the grief. Letting the grief pass on is not about forgetting the one you grieve over. You never forget but you can stop the grief and get on with your life.

We have seen that early man had situations that produced great stress. Our bodies react to this and hopefully help us survive the event. Then it's gone. In modern society our stress usually never ends and this is harmful to us. Grief is a form of stress. It serves a purpose and then is gone. When we hold onto to it for too long, it becomes dangerous to our health.

### **Changing Life Styles**

You MUST change your life style to lower your stress. Your body is screaming at you to slow down. It can't take the hectic pace. If you keep it up, your body will get ill to force you to slow down. Do it, before you have no choice! Slow down your life. Spend more time on things that will really matter on your death bed. One rule we find to be so important in life is "DON'T WORRY ABOUT THINGS YOU CAN NOT CHANGE"! This is easier said than done for most of us. Over our life we build up worries until almost everything becomes stressful. Do you worry about?
- Driving at night

- Having an accident
- Plane crashes
- Being mugged
- Getting sick
- What others think about you
- Being accepted

These are not things you should worry about. The truth is there is very little you can do about them. Live your life as stress free as possible and enjoy it more.

As parents we can pass these traits on to our kids and it becomes part of their life as well. This may happen because they got both our genes and epigenes or because they observe us and learn from us or a combination of the two. Do your kids a favor for life and teach them to relax and enjoy their lives.

## *Science*

According to lead researcher Mark Wilson, PhD, chief of the Division of Psychobiology at Yerkes, "Chronic stress can lead to a number of behavioral changes and physical health problems, including anxiety, depression and infertility."[192]

## *Conclusion*

Stress is a major factor, along with diet, affecting our health and well being. We can't always eliminate stress totally but we can do a lot to reduce it. Always ask yourself, "Is there anything I can do to change what I am worried about?" If the answer is YES, do it! If the answer is NO, forget about it! Why stress over something you have no control over?

# Chapter 17:

# Conclusion

## *Introduction*

We have shown you many examples of work being done that show the affects of diet, thinking, belief and stress on our health. You as the reader are tasked to do your own research. Remember to ask who benefits from whatever message you hear? You and only you are in charge of your health. Don't rely on anyone else. Do your research and make your own mind up. It may save your life and provide you with a healthier and happier existence.

## *Basics*

### It's your Health and Life at Risk

John Assaraf says: "Here's the problem. Most people are thinking about what they don't want, and they're wondering why it shows up over and over again."[193] If we think over and over again that we are going to get sick . . . we will! Our brains are very powerful and can resolve many of our problems along with the help of our DNA and epigenetics. JUST USE THEM FOR THE GOOD!

Humor is an ability of humans that seems to not have a purpose or does it? "The power of humor gives those with it an advantage over those that do not have it."[194] Humor also gives us a better ability to think positive and over look our negative feelings.

Buddha wrote:

> *Don't chase after the past,*
> *Don't seek the future.*
> *The past is gone.*
> *The future hasn't come yet.*
> *See clearly on the spot*
> *That object which is now.*

As we said above, get started now. What you may have done or not done in the past is not important. Don't dream about the future, it is not yet here. If you really want to change your life, your health, your happiness . . . Do it today. Each of us makes our future by the decisions we make in the present.

In the bestselling book **Eating Well for Optimal Health**, Andrew Weil, M.D., outlined his seven basic principles of diet and health:

- We have to eat to live.
- Eating is a major source of pleasure.
- Food that is healthy and food that gives pleasure are not mutually exclusive.
- Eating is an important focus of social interaction.
- How we eat reflects and defines our personal and cultural identity.
- How we eat is one determinant of health.
- Changing how we eat is one strategy for managing disease and restoring health.[195]

No one really knows everything about our health or the diseases that plague us but isn't it better to trust God and His natural solutions that have evolved over the millions of years, rather than man's solutions which at best are 200 years old and most much less than that. We don't know about you, but we put our faith in God.

Finally Dr. Ornish, MD in his book, The Spectrum, writes:

> "The number one cause of death in most of the world is almost completely preventable just by changing diet and lifestyle."[196]

This is the heart of my book! Work on your diet, do some exercise and meditate to reduce stress. You will reap the benefits of a healthier life and a happier one. If you are already ill or dying, make these changes quickly, it can save your life.

**Top Ideas to Remember from This Book**

1. Epigenetics puts our health and happiness in our hands
2. We all have roughly the same DNA
3. Our epigenetics is very changeable
4. Only you can make your life and health better
5. Yes we are all going to die but the important question is will you have a good quality of life during your last 10-20 years on Earth
6. Eat less calories in more small meals and live longer
7. Reduce the amount of animal protein you ingest
8. Sugar is addictive—reduce the amount you eat or eliminate it all together
9. Fruit juice is NOT fruit—avoid it
10. Our diet has changed drastically since World War II and our health has gotten worse
11. If you still smoke—STOP STOP STOP
12. Sugars in fruits are better than refined sugars
13. Milk is for cows—NOT humans Don't drink it
14. Fruits & vegetables are good for us—Eat more of them—Mix them up
15. Vegetarians are healthier
16. Beware of TV ads they are designed to make you buy what you don't need or want, they are really designed to make someone else a lot of money
17. America is getting sicker and it is costing us more in our quality of life and our tax dollars
18. Drug companies are in the business of making money not curing your diseases
19. Most modern diseases can be cured by proper diet and reduction in stress

20. Reduce your stress
21. Get more exercise
22. To lose weight you must eat—eat many small meals—skipping meals will turn on the "Starvation Mode" and store excess calories as fat.
23. We are what we eat

**Top Myths to Change in our Lives**

1. You must eat meat protein to be healthy
2. Protein is from meat only
3. Protein is good for you; eat all you want
4. Vegetarians are missing critical parts of a healthy diet
5. Food has no effect on health
6. Some of us have good DNA; others have bad DNA (In reality we have about the same DNA)
7. Health is determined by our DNA alone ( it's not it's from our genes and epigenes in combination)
8. I have to take a lot of pills to stay healthy
9. I don't have time to cook

**Future Directions**

Epigenetics and nutrigenomics will spawn new research into new drugs. It will map which foods do what to cause or cure certain diseases. We may even see genetic counselors that will look at our epigenetic make up and recommend a diet. These will happen because someone will make money doing them. In reality we evolved with certain food groups. The closer we stay to these foods and ways of thinking the better off we will be. The more we move to modern ways of eating and thinking the sicker we will become. You don't need new drugs! You don't need genetic counselors! You don't need new foods! Everything you need is here and has been for millions of years. All you need to do is choose it!

## *Losing Weight by Eating Healthy*

Try eating 5 small meals per day instead of 3 larger ones. If you eat breakfast between 7 and 8 am then target a morning snack around 10am and lunch between 12 and 1 pm. The afternoon snack should be around

3pm and dinner between 6 and 7pm. If you don't feel hungry for your two snacks, you are eating too much at the meal times. Cut back on some food so you are hungry before each of the 5 meals. Eating 5 instead of 3 meals lets our bodies know that food is readily available. This will prevent you from going into "starvation mode". Starvation mode was an old mechanism that prevents the body from easily giving up its fat when it thinks food is scarce. This would have helped early man as they roamed the land looking for food. Today it is of no use since food is readily available to most of us.

We have a natural feeling that skipping meals will help us lose weight since we are not eating any calories. This can't be farther from the truth. Skipping meals will turn on starvation mode and your body will fight you for every pound of fat it gives up. Let your body relax in knowing that food is plentiful by eating smaller meals. Your metabolism will increase and your body will work with you to shed the fat pounds off.

Diets should not be viewed as restrictive. If you do it won't work. Mentally adjust yourself that this is how you are going to eat for the rest of your life. It will not only rid you of excess body fat but make you healthier and happier. I have compiled a series of dos and don'ts for you:

## Dos

- Eat more plant-based foods (fruits and vegetables). These foods will slim you down and make you healthier. If you have an illness they may even help cure it or diminish its affects.
- Eat a wide variety of different plant-based foods. Eat things with different colors (yellows, reds, blues, etc.).
- Find recipes for making your foods that excite you. If you enjoy eating it, you will continue for a life time.
- If you have been out of the kitchen; get back into it! It can be fun.
- Eat 5 smaller meals per day.
- Reduce the amount of meat and animal based foods in your diet (milk, meat, eggs, cheese, etc.). If you can eliminate meat and only eat a small amount of fish the better. Being a total vegan is best. Even if you must eat some meat, try to keep the calories from all sources of animal protein (milk, meat, cheese, eggs, etc.) to less than 10% of your total caloric intake.
- Feeling hungry before each of the 5 meals is a good thing. Your metabolism is increasing.

- Get to know where foods you eat are on the Glycemic Index. The Glycemic Load is even better. It takes into account not only the Glycemic Index number but the calories as well. Carrots are relatively high (for vegetables) on the Glycemic Index but low in calories making them a good food. You would have to eat a ton of them to raise your blood sugar level rapidly.
- Drink plenty of water all day long. As a rule I use ½ my weight in ounces as a daily guideline. For example, if you weigh 160lbs, you should drink 80 ounces of water which is about 10 cups of water.
- Eat nuts like almonds or pecans. Buy them in bulk, raw if you can. I find a lot of stores have them but they are not always fresh. If you can't get fresh raw nuts buy unsalted roasted nuts.
- Reduce the amount of salt in your diet by reducing the use of packaged goods (cans, frozen, boxes). We get a lot more salt from factory produced foods than from a salt shaker.
- Space meals apart by 2 to 2.5 hours minimal. This allows your body to finish digesting the previous foods and be ready for new food.

**DON'Ts**

- Never eat within 2 hours of going to bed. It will upset your digestion process.
- Never drink or eat too much. This may get you sick.
- Don't eat fried foods or greasy foods of any kind. Your body will adjust its taste after a while so that you don't miss them or even like them. Fruits and vegetables will begin to taste better than they ever did.
- Never supersize anything! Supersizings will supersize you back.
- Reduce or eliminate sugars. Sweets are the enemy. Eating things that high on the Glycemic Index will make you fat and could lead to diabetes. Remember high things on this index are not only deserts but cereals (except oatmeal), white breads, muffins, pancakes, beer, etc. Some carbohydrates are good (low on index) and some are bad (high on index).
- Learn yoga and meditation. These techniques will keep you from being depressed. Depression and sadness are your enemies. They will lead you to binge eating, a bigger waist and illness.

- Reduce and eliminate coffee intake. It will cause your body to fight you on each pound of fat lost.
- If you are trying to lose weight, reduce the amount of bread and pasta in your diet. Make them special not everyday items. Always consume multi grain or whole grain types, never white processed versions.

## Sample Daily Menu

Here is a sample of the way we eat:

### *All Day*

Drink plenty of water all day long. You cannot lose fat without drinking water.

### *Breakfast*

We usually eat oatmeal (choose an oatmeal that is all grain, not instant and has a lot of fiber but no sugars added) with a banana and berries mixed into it. I do not use any milk or sugar. Your taste will adjust to the non-sweet flavors. I have a cup of hot green jasmine tea with it. Coffee should be reduced or avoided (in some tests caffeine has been associated with increased body mass).

### *Morning Snack*

We usually have a piece of fruit or a handful of nuts (unsalted raw almonds or raw pecans).

### *Lunch*

We usually have 1 or 2 pieces of fruit for lunch or some vegetables or homemade soup. Canned soup is VERY high in salt. Homemade soups are easy to make.

### *Afternoon Snack*

We may have a piece of fruit or another handful of nuts.

## *Dinner*

We usually make a soup or a plate of different vegetables depending on what is in season. Occasionally we will have fish or pasta. Again they are the exception not the norm. A nice salad with a variety of vegetables tastes good and is healthy. Watch the oils! We use mostly balsamic vinegar with little or no oil.

## *Evening Snack*

If you feel very hungry two hours after dinner and at least 2 hours before bedtime, a sixth optional meal is OK. Have fruit! It will satisfy your sweet tooth and hunger.

## Meat Substitute Products

We occasionally will use a meatless food product like ground beef substitute, sausage substitute, "lunch meat" or veggie-burgers. They are satisfying when done correctly and fulfill any lingering desire for meat.

## Useful Cookbooks

We have also used several cookbooks but generally enjoy experimenting on our own with different foods and spices. Everyone needs inspiration so here are some books to look at:
1. **The Spectrum** by Dean Ornish, MD—This book has information on staying healthy or curing a health problem followed by recipes in the rear of the book.
2. **Cooking the Whole Foods Way** by Christina Pirello—As seen on public TV—A book on macrobiotic diets that are totally vegan and delicious.
3. **Skinny Bitch Ultimate Everyday Cookbook** by Kim Barnouin—Delicious and healthy recipes that will help you lose weight and keep it off and stay healthy.

## *Final Thoughts*

People always ask us "What do you eat?" We picture them thinking of us eating beans, nuts and berries. While we do eat these items, we also eat old favorites that are "non-evil" versions, like spaghetti and meatless meatballs, sausage (meatless) and onions and peppers on a great roll, Meatless meatloaf and many more.

Diet, exercise and life style all play a part in being healthy and happy. We have not said much about exercise in this book but we believe it is very important. After all, early man and woman were nomadic and probably walked all day in search of food and water. Exercise alone will not keep you healthy. We hear all the time of famous sports figures that die of cancer and heart attacks. They were in shape and exercised a lot. Americans are obsessed with weight. Many feel all that is important is being in shape and looking good. While being in shape is a noble goal, it can provide you with a shapely body but does not guarantee a healthy one.

**Remember if your answer is "I don't have the time to do healthy things" you will when you get sick.**

We hope we have left you with some food for thought. Don't just put the book down and say interesting. Make a list of things you need to change in your life today. Get help if needed. Start making changes and reap your new health. Above all be happy and feel blessed. **It's your decision!**

The ancient Roman stoic, Seneca, Socrates, Plato and Aristotle all agreed "to live is not a blessing, but to live well."[197] Alexandra Stoddard's book, You are Your Choices", talks about each of us being the summation of all our choices. Epigenetics is the biological method by which this happens. Each of us has different events, decisions and interfaces to the world around us. Some of these experiences are good and others are bad. They all *kick-off* epigenetic reactions that in turn express and suppress genes in our cells. We are the summation of all these expressed and suppressed genes. We are unique.

If you are not smiling more, laughing more, eating healthy, forgiving people more, loving all life more and enjoying the creativity of your wonderful mind; then you are doing yourself a grave injustice.

Buddha said, ***"We do not generally suffer from others; we suffer from ourselves. I have shown you the way to liberation, now you must take it for yourself."***

"It is common sense to take a method and try it. If it fails, admit it and try another. But above all, try something."—Franklin Delano Roosevelt[198]

We hope you at least have some familiarity with the word epigenetics after reading this book. Hopefully, you will also have a desire to learn more about it. This science is changing drastically as we write this book. Search epigenetics often and keep up with the latest ideas.

We hope you will evaluate your life and answer the question: "Is it working for you?" If the answer is no; admit it to yourself and try something else! The most important concept from this book is: **"Do your own research, learn more about these topics and change your life. Only you can change your life for the better."**

For the religious out there, we believe God gave us this ability for change to make our lives better or worse. It is our decision (free will). He also gave us things to help improve our lives. He gave us food that can repair us and make us healthy, clean water and air for us to survive in, and faith to help us believe in good. It is your choice. Choose well!

We hope that our scientists will have the wisdom to use the new knowledge they discover to help all life forms on Earth; to make our lives healthier and happier. We also hope that they will not just think in terms of new drugs to be sold. Hopefully they will follow scientific methods and seek the causes not just the symptoms of diseases. Let us not just fix the symptom but eliminate the causes and the diseases themselves.

# Appendix A:

# Recommended Books, Videos & Web Sites

## Books & Articles

There are so many great books on this subject as well as medical papers, websites, etc. The following were publications that changed our lives and our way of thinking.

1) T. Colin Campbell, PHD. "The China Study" 2005
2) Dean Ornish, MD. "The Spectrum" 2008
3) Fuhrman, Joel, M.D. "Eat To Live". 2003
4) Watters, Ethan. "DNA is not Destiny". Published online
5) Shenk, David. 'The Genius in all of us". Doubleday, 2010.
6) Khalsa, MD, Dharma Singh. "Food as Medicine". Atria Books.
7) Weil, MD, Andrew. "Eating Well For Optimum Health". Knopf.2000
8) Editors of FC&A Medical Publishing. *"Eat and Heal"*. FC&A Medical Publishing, 2001.
9) Stoddard, Alexandra. "You are your choices". Collins. 2007
10) Dobson, Roger. "Death Cab Be Cured". MJF Books
11) Dr. Lipton, Bruce H. "The Biology of Believe: Unleashing The Power Of Consciousness, Matter And Miracles" Hay House Inc. September, 2008
12) Dr. Bruce Lipton. The Magnetic Centre. "The Biology of Belief: An Epigenetic Primer".
13) Moll, Rob. "The Art of Dying". IVP Books. 2010

14) Robbins, John. "Diet for a New America".Pg 177
15) Chopra, Deepak. "Journey into Healing". Crown Publishing Group. March 1995
16) Bonnie Minsky MA, MPH, LDN, CNS & Steve Minsky Nutritional Concepts, Inc. "Using epigenetics to prevent chronic disease: Part One". 2007.
17) Flanagan, M>S>S>W>, Beverly. "Forgiving the Unforgiveable". Macmillan. 1992
18) Church, Dawson. "The Genie in your Genes". Energy Psychology Press & Elite Books. 2007
19) FC&A Medical Publishing. "Eat and Heal". FCA. 2001
20) Nicholas Wade. New York Times. Aug 2, 2010. "Breast milk sugars give infants a protective coat"
21) Adams, Chris. "Student Doctors Start to Rebel Against Drug Makers' Influence". Wall Street Journal June 24, 2002
22) Nicole M. Avena, Pedro Rada, and Bartley G. Hoebel. "Evidence for sugar addiction: Behavioral and neurochemical effects of intermittent, excessive sugar intake". National Institute of Health
23) Perkins, M Ed., Cynthia. "The hidden dangers of sugar addiction". No-Hype Holistic Health Solutions.

## *Cookbooks*

1. Dean Ornish, MD. "The Spectrum" 2008
2. Kim Barnouin. "Skinny Bitch—Ultimate Everyday Cookbook" 2010
3. Christina Pirello. "Cooking The Whole Foods Way" 2007

## *Videos on YouTube*

1. Dr. Bruce Lipton—"Epigenetics: How Does It Work"
2. "The Gene Expression".
3. "The Epigenome at a Glance" from Learn Genetics.
4. Dr. Bruce Lipton—"Epigenetics: Your unlimited potential for health"
5. Barnett, Dr. Matt—"The Epigenome Song"
6. A must watch set of videos is by Dr Michael Klaper MD explains how eating animal flesh and fat can make us ill. Here is a list of six parts on YouTube:

a. http://www.youtube.com/watch?v=TF2MZN6ImB0
b. http://www.youtube.com/watch?v=Zn-7sfxSmWc
c. http://www.youtube.com/watch?v=41Y2qnE373k
d. http://www.youtube.com/watch?v=dyB79viq74w
e. http://www.youtube.com/watch?v=K4C-2Um4HzM
f. http://www.youtube.com/watch?v=029f9JppOtA

## *Web Sites*

1. *www.rainbowblessings.org* Rainbow blessings website is dedicated to anyone dealing with a terminal illness or the people that are dealing with their lives being changed because someone they love has a terminal illness. It is written about Gail's story (Jo Anne's first husband who died of ALS). Please watch Gail's videos on this site. They will move you and may help you deal with your own problems.
2. Furuya, Yukio. "The Relation of Effects of Dietary Changes to Physical and Mental Disorders and Crime Occurrence among Youths". *http://www.jicef.or.jp/wahec/ful418.htm*
3. The Chicago Western Electric Study. "Relation of vegetables, fruit, and meat intake to 7-year Blood Pressure Study in middle aged men. *http://aje.oxfordjournals.org/cgi/reprint/159/6/572.pdf*
4. Vegan Diet Helps Fight Prostate Cancer, Study Says. *http://www.vegtaste.com/pages/posting.php?articleId=200*
5. Hirayama, T., "Epidemiology of Breast Cancer with Special Reference to the Role of Diet, *Preventative Medicine 7* (1978): 173-95
6. "Dairy Products Linked to Prostate Cancer," *Associated Press,* April 5, 2000
7. Ophir O., et al., "Low Blood Pressure in Vegetarians . . .," *American Journal of Clinical Nutrition* 37 (1983):755-62.
8. Key, T., et al., "Prevalence of Obesity Is Low in People Who Do Not Eat Meat," *British Medical* Journal 313 (1996):816-7
9. Halweil, Brian, "United States Leads World Meat Stampede," Worldwide Issues Paper, July 2, 1998.
10. Serge Ahmed, Ph.D. "Intense Sweetness Surpasses Cocaine Reward" *www.plosone.org/article/info:doi/10.1371/journal.pone.0000698*

11. Bart Hoebel, Ph.D. "Sugar can be addictive, Princeton scientist says." *http://www.princeton.edu/main/news/archive/S22/88/56G31/index.xml?section=topstories*
12. Serge Ahmed, Ph.D. "Intense Sweetness Surpasses Cocaine Reward" *www.plosone.org/article/info:doi/10.1371/journal.pone.0000698*
13. familydoctor.org. "Added Sugar: What You Need To Know". *http://familydoctor.org/online/famdocen/home/healthy/food/general-nutrition/1005.html*
14. Baldauf, Sarah. "Foods Surprisingly High in Added Sugar". Yahoo Health. *http://health.yahoo.com/featured/35/foods-surprisingly-high-in-added-sugar/*
15. Phillips RL. Role of lifestyle and dietary habits in risk of cancer among Seventh-day Adventists. Cancer Res 1975;35(Suppl):3513-22.
16. Rory Hafford. Irish Medical Times. "Diagnosis and management of cow's milk protein allergy". *http://www.imt.ie/clinical/paediatrics/diagnosis-and-management-of-co.html*
17. PubMed.gov. "Is type 1 diabetes a disease of the gut immune system triggered by cow's milk insulin?" *http://www.ncbi.nlm.nih.gov/pubmed/16137120?dopt=Abstract*
18. Lauren Cox. Live Science. "Cigarette smoke jolts hundreds of genes, researchers say". Thu Jul 15, 9:55 am ET. *www.livescience.com*
19. THE BIOLOGY OF BELIEF: An Epigenetic Primer. *http://www.danbartlett.co.uk/lipton_epigenetics.htm*

## The Epigenetics Project BLOG

Remember to keep up to date on these concepts by visiting our blog at *http://georgefebish.wordpress.com*.

# Appendix B:
# Thoughts to Remember

1) It's not about dying but how we live the last 10 or 20 years of our life that really counts. This is our quality of life. Our society with its pills is causing us to live a little longer with our diseases but our quality of life is poor.
2) The problem is that drugs are fixing nothing. We are getting fatter, feeling more depressed and getting sicker. The only thing that these actions do is to make certain individuals and corporations very wealthy.
3) The reason we don't see ads on healthy ways to live is no one makes any money by keeping us healthy, thin and happy.
4) Imagine an America that eats right, feels good and is not obese. The health care system with all our hospitals and doctors would collapse. No need for the number we currently have. The drug companies would go out of business. The diet companies would as well. Our taxes to cover the huge cost of America's illnesses would decrease drastically. Our politicians would have to find something else to spend their time on. This is not a pipe dream. We can obtain this but we won't if we continue on our current road.
5) Does anyone believe we are really working on a solution to cancer or AIDS or Diabetes? No way! Can you imagine the revenue loss in drugs if they suddenly cured it?
6) The cell doesn't know what is good or bad. It reacts to changes in its environment. The signals you send to your cells is all that counts.

7) Do we really believe deep down, that natures natural foods that we have been eating from the beginning of time is not as good for us as man's bottled products that are 20 or 30 years old or less?
8) With every feeling and thought, in every instant, you are performing epigenetic engineering on your own cells. The question is "Are you a good or bad engineer?"
9) Listen better to what our bodies are telling us, after all our bodies have been doing this for millions of years.
10) Who we are is really a summation of the choices we made in our lives. Death ends the possibility of making choices but while you are alive you can still choose. Choose wisely.
11) Millions of people have embraced vegetarianism in one form or another.
12) Imagine how powerful you are, you can change your life by changing your food.
13) *"How do you get your protein?"* People are dumb struck when they learn that protein is everywhere. Every living thing on Earth is made of protein.
14) The power of marketing is to sell us stuff even if we don't need or want it.

# Appendix C:

# Datamation Article in Paradigm Shift Column

## *Does DNA Use Remote COM?*

**By David E. Y. Sarna and George J. Febish**

We often use existing real world models to understand software design problems . . . recently an article on polymorphism an encapsulation appeared in a Biology book . . . talk about a Paradigm shift.

George often tells the story of late night discussions with his brother-in-law Nick, a Doctor of Biology studying DNA, on the subject of DNA vs. Software programming. Nick once told him about a problem when DNA is transcribed into RNA to build protein (set of amino acids which are the building blocks of proteins) needed by the body. If a particular code (set of amino acids), TGA, was not at the end of the chain (it signals stop to the ribosome machinery) it would randomly continue through the cell replicating garbage (garbage in garbage out). George's answer was a missing end of record indicator. This obviously over simplifies the structure of our DNA but it makes a good point that software is a great model for Biologists and DNA is great model of a working distributed object orient system for software designers.

Biologists have discovered that tissue-specific genes are turned on in specific tissues at specific times in development by other proteins, the products of other genes. This is not unlike software setting values at one point that effect the logic flow later in the program. Differentially expressed genes are controlled by DNA sequence elements in front of the gene ("objects") that are relatively small and can be easily introduced in different contexts to achieve the same output on other genes. This in software would be known as polymorphism.

Nature had a few more years to work on a more elegant solution than most of us have. Nature posse's two qualities we developers seldom adequately design into our applications: Adaptability and reasonable error handling. It can be useful to study this model and use it to see where we might go with our current object models. Clearly nature favors smaller pieces that make up complex functionality. The pieces are not intelligent but the sum of the parts is. We programmers, on the other hand, have always favored enormously large applications that lack reusable functionality. Why do you think that's true? Every science and engineering discipline discover small desecrate components are better. Are we truly brain dead or just stubborn?

Imagine a software application made up of thousands of reusable parts, each one modifying how it operates. Those components that work well will tend to survive and inherit their new methods to other objects. Those that fail will be eliminated forever more. Current software theory tends to think of zero defect code, objects that are hard-coded by their programmers and never change. But imagine software that used bugs to modify itself, much as our bodies do, to produce better and worse alternatives. The worse alternatives would be eliminated while the better ones would be propagated into future generations. This might be a first step towards *thinking* software.

DNA obviously has inheritance but can also be used as a model for the COM/ActiveX school of thought. The religious wars of aggregation vs. Inheritance can take note that nature did not implement inheritance ALA OOP. Can you imagine if we continued to dynamically inherit good and bad things from our parents as they changed? It's bad enough getting a onetime snapshot. We need a firewall to protect us, aggregation provides such protection. One might think of aggregation as safe inheritance.

Biologists are dividing DNA into individual triplets to categorize the DNA codes. Some Biologists feel that once this is done they will truly understand DNA and its workings. We feel this is only the first step. Imagine an alien landing on Earth and studying a PC. After years of research they final crack the machine code! The Intel instruction set. They know everything about that PC, or do they. Can they easily infer from the instruction set or a listing of instructions from memory that this particular

PC was running an accounting package under Windows/95? The concept of software is much more elusive than hardware. In fact can anyone really go back from the low-level codes to the thought that created them? We have disassemblers that take machine codes and translate them into a more usable language like assembler or C but that doesn't tell us what the code was doing from 3,000 feet.

DNA seems to be more like hardware OP codes, could there be another layer, an invisible layer of logic we call software on top of DNA? How does it get programmed? If DNA is the machine op codes of our body what is the Operating System? Where are the programs stored? Where is data stored? How do they communicate with each other?

Is a set of DNA instructions used to build an RNA strand to then build some protein, similar to an application instantiating an object to do something on its behalf? Both are code based, both have an interpreter for the codes, both are built up of many desecrate components. In fact parts of our DNA are the same in all mammals indicating evolution kept objects that worked well and got rid of those that did not. Will a future repository decide on its own which objects stay and which go? Will lawyers sue its creator for discrimination? We can just see a future F. Lee Bailey winning a case that decides what objects can and can't do. Object rights the call of the year 2,000. And we thought we understood the year 2,000 problem!

Our bodies react to many complex signals and a lot of logic is processed and I don't mean to over simplify it. Think of what happens when you accidentally touch a very hot surface. Scientists tell us an involuntary reaction quickly moves the hand away and then tells the brain HOT. Could this involuntary function be a like a remote COM object that acts on its own and communicates back to the higher layer only when needed?

We could even make a case for a multi-tier architecture at work here. Imagine that our senses are like an application's GUI. These are the high level functions that drive us all. The middle layer consists of processes in the brain and throughout the body that interact in a network of rules, logic, body functions, controls, etc. The lowest layer is the encapsulation of knowledge in the brain (our database).

How far can we take the analogy? Are we just having fun or is there something to this ranting? We are getting too much into the biology side let's get back to what we do understand, Object Oriented Software Systems.

The ideal situation for corporate America is to start heading in this direction. Imagine systems that are built up with components interacting with each other. Some of these components have authority to make decisions and act on their own, like the involuntary mussels, while others need to first check with their supervisor (a module higher up in the tier). Knowledge is converted to program language only once and reused in any other applications requiring it.

Like our human brain, which has logic built upon our more animalistic functions, the sea of objects would start small and evolve into more mature and sophisticated functionality. Smart people not bogged down by the techniques of programming can easily create at this point new corporate logic and functionality.

The corporate network will expand beyond corporate boundaries to intra-corporate data much as we humans expanded beyond our own limits by communicating with other people. Communication is the key to future language. Communication requires a common language by which the communication occurs. I believe this will be a derivative of COM that allows any computer system to communicate with another. Sound waves carry human communications, the Internet will carry computer communications.

The human body and its evolution seems the perfect model of computer logic, remote communications, and networking. Does it hold secrets for us future thinkers of computer technology? Can we both, biologists and computer nerds, learn together from each other's experience and knowledge. What does Bill Gates know about DNA that we don't? Why is Microsoft in the genetic business? Will future releases of NT give birth to something really new? We know it will be painful. Stay tuned who knows . . . but beyond 2,000 will be different. For one thing we won't be talking about the year 2,000 problem.

# Appendix D:
# Powerful Ideas in this Book to Remember

## *Introduction*

Epigenetics and genetics are a memory system. *Bad ideas die and are forgotten. Good ideas are remembered and passed on to future generations.*

## **Chapter 1: Early Life, Early Man and DNA**

Food is usually available even though our bodies may think it is not. If you want to lose weight fast, the most obvious answer would be not to eat! This would be the wrong choice. It would enable that ancient "Starvation Mode" mechanism and our bodies would fight us to protect each pound of fat. If we were to eat small meals of fruit, vegetables and nuts several times a day, we would not be hungry and the "Starvation Mode" mechanism would switch off causing excess calories to be excreted as waste and any stored fat to be burned off. So to lose weight, you must eat! You must eat smart not stupidly and that is the trick!

Keeping with our computer analogy; some functions of genes are old and may be dormant or not needed anymore. These genes are not normally turned on but the data we spoke about above can affect them and turn them on. Your computer may have a lot of programs. Some may be really old and may no longer work. They cause no damage as long as they are not activated. Activate them and anything can happen. We now know that some genes can cause cancer but only if they are activated!

## Chapter 2: Epigenetics

How we live our lives, become the music our epigenetics plays on our genes. What kind of music are you playing?

We have all been programmed in our society to be a victim. Therefore we are victims.

Self healing works! Why can't we use it better? Most of us have bought into the idea of our genes control us. Some of us have good ones and others bad ones. Modern Biology shows us this is totally false. Our genes do nothing until they are turned on (expressed). Epigenetic factors express our genes.

## Chapter 3: Nutrigenomics

Cancers have been observed to have high levels of methylation (epigenetics). We now know that foods as well can aid in the prevention of cancer. Is there a link between certain foods and cancers?[199] These studies suggest that there are GOOD and BAD foods. Bad ones include: Red Meat, processed meat, grilled meat, dairy, animal fat, partially hydrogenated fats, etc. Good foods include: fish, fruits, vegetables, tree nuts, omega-3 fatty acids, whole grains, etc.

Cancer, once believed to be caused by mutations of our genes, is now believed to be caused by genes that are turned on via epigenetics.

## Chapter 4: Environment—You are what you Eat, Smoke and Drink

Our diet and our levels of stress can cause these reactions. Our bodies are a multitude of living organized and intelligent cells working together and communicating with each other to determine what is best for the organism and the continuance of the DNA code.[200] Even Plato recognized the importance of diet and stress.

> *"And we have made of ourselves living cesspools, and driven doctors to invent names for our disease"*—Plato

We like to think about what we are eating and ask the question: "Did early man eat this way?" If the answer is Yes, it is fine to eat but if the answer is no; avoid it with a passion.

## Chapter 5: Environment—You are what you Think

Stress can cause us to think negative and that kind of thinking can and will cause us to be sick. Our mind is such a powerful tool that in the wrong hands, it actually will hurt us.

Thinking is a great gift given to man. It also is a great curse if used wrong. Thinking over and over about something you can't change, can raise your blood pressure and harm you. Meditation is an exercise that can help break this cycle of thinking and bring peace to our inner soul. Use your gift of thinking wisely!

## Chapter 6: Environment—You are what you Believe

Who we are is really a summation of the choices we made in our lives. Death ends the possibility of making choices but while you are alive you can still choose. Choose wisely.

## Chapter 7: Vegetarianism

Imagine how powerful you are, you can change your life by changing the amount of plant based foods in your diet.

If you are going through cancer or heart disease or know someone that is; we highly recommend you give them a gift of this work. It could save their life. There is no downside. Eating vegetables will NOT kill you.

A minimum of $60,000,000,000 in annual medical costs in the United States is directly attributable to meat consumption. Compare that to $65,000,000,000 in annual medical costs directly attributable to smoking.[201] We all see the risks of smoking but why can't we see the risks of meat consumption?

## Chapter 8: Glycemic Index

Modern diets often have us eating foods that are full of sugars or break down into sugars with a high GI (see Figure 19: Glycemic Index Table) These quick rises can cause cravings to eat even more. The body tries to respond to the high amounts of sugar with insulin (required for cells to be able to absorb the sugar as energy). If this happens often, we stress the cells of the body causing them to resist the insulin (Type 2 diabetes).

## Chapter 9: Take Control of your Thoughts, Actions and Diet

Good marketing companies get paid the big bucks to fool us. If they can sell you a bad product or negative idea and make it seem like it is good for you, they have done their jobs well. The worst part is most Americans sit in front of the TV all evening. They are exposed to more and more commercials. The concepts in these commercials become facts. If you hear something over and over again, you believe it. The Internet is even worse. Pseudo-facts are published all the time. If we read it online we tend to believe it. Do your own research! Draw your own conclusions! Live YOUR life well!

## Chapter 10: Don't let TV Advertisements Control your Actions

Americans consume TV like they do sugar. Marketing companies know this. They target us with ads that we believe because we have heard them so often. What you hear in an ad is not necessarily for your benefit. It is to sell products and make money. Lots of money is made through the use of clever ads.

## Chapter 11: America is Sick, Literally

It is very clear that we are a nation of sick people and we are getting sicker. Modern medicine is unable to stop it or even slow it down. For all the drugs we take, nothing is working. Do your own research. Think about what is really happening with health in America. Be one to stand up and stand out as an individual who will not allow this to happen to you. Change your life; you only have one shot at it.

## Chapter 12: Drugs—Pro & Con

We have become a drug nation. If you believe the commercials, pills will cure all. Some of us need some pills to fight an infection or a disease, but many pills are unnecessary. Many people end up taking pills to offset affects of other pills.

## *Chapter 13: Sugar addiction*

Do we need to have an addictive substance added to our foods to like them? Are you starting to see the differences between artificial commercial foods and nature provided real ones? Nature has packaged food to provide us with everything we need to be happy and healthy. The packaging was worked out over millions of years. We evolved with these food supplies. They are natural to us. Commercial foods are at best a few hundred years old and the most destructive are only about 50 years old. The purpose is not to nourish us but to addict us and make money from us. Are you paying to be unhealthy?

## *Chapter 14: How do you get your Protein*

Proteins come from ALL foods (plant and animal based). Animals get their protein from plants. When we eat meat, we are getting proteins from plant based sources that the animal ate or another animal eaten that ate plant based foods.

We believe foods closer to us on the evolutionary chain can cause more harm than foods farther down the chain. Plant-based foods are much farther away from us (evolutionarily) than animal-based foods are.

Vegetarian diets are extremely healthy and provide all the nutrients and antioxidants nature made for us.

## *Chapter 15: Diseases and Cures*

A bad habit such as smoking can cause a drastic epigenetic change in our bodies resulting in illness and death.

Some people have said to me that eating healthy and organic is too expensive. They cannot afford *it*. We always tell them you have a choice. **Pay now with healthy foods or later with drugs and medical procedures. The choice is yours.**

## *Chapter 16: Epigenetics—Stress*

Stress is a major factor, along with diet, affecting our health and well being. We can't always eliminate stress totally but we can do a lot to reduce it. Always ask yourself, "Is there anything I can do to change what I am worried about?" If the answer is YES, do it! If the answer is NO, forget about it! Why stress over something you have no control over?

## Chapter 17: Conclusion

John Assaraf says: "Here's the problem. Most people are thinking about what they don't want, and they're wondering why it shows up over and over again."[202] If we think over and over again that we are going to get sick . . . we will! Our brains are very powerful and can resolve many of our problems along with the help of our DNA and epigenetics. JUST USE THEM FOR THE GOOD!

Dr. Ornish, MD in his book, The Spectrum, writes about heart disease:

> "The number one cause of death in most of the world is almost completely preventable just by changing diet and lifestyle."[203]

## The BLOG

Remember to keep up to date on these concepts by visiting our blog at *http://georgefebish.wordpress.com*.

# Appendix E:

# Top Ten Diseases

## *Men's top 10 Disease Preventions*
By Mayo Clinic staff[204]

Do you know the greatest threats to men's health? The list is surprisingly short—and prevention pays off. Consider this top 10 list of men's health threats, compiled from statistics provided by the Centers for Disease Control and Prevention (CDC) and other leading organizations. Then take steps to promote men's health and reduce your risks.

### No.1—Heart disease

Heart disease is a leading men's health threat. Take charge of heart health by making healthier lifestyle choices. For example:
- Don't smoke or use other tobacco products. Avoid exposure to secondhand smoke.
- Eat a healthy diet rich in vegetables, fruits, whole grains, fiber and fish. Cut back on foods high in saturated fat (animal protein e.g. meat, cheese, eggs and milk) and sodium.
- If you have high cholesterol or high blood pressure, follow your doctor's treatment recommendations.
- Include physical activity in your daily routine.
- Maintain a healthy weight.
- If you choose to drink alcohol, do so only in moderation. Too much alcohol can raise blood pressure.
- If you have diabetes, keep your blood sugar under control.
- Manage stress.

## No.2—Cancer

Lung cancer is the leading cause of cancer deaths among men—mostly due to cigarette smoking, according to the American Cancer Society. Lung cancer is followed by prostate cancer and colorectal cancer. To prevent cancer:
- Don't smoke or use other tobacco products. Avoid exposure to secondhand smoke.
- Include physical activity in your daily routine.
- Maintain a healthy weight.
- Eat a healthy diet rich in fruits and vegetables, and avoid high-fat foods (animal protein e.g. meat, cheese, eggs and milk).
- Limit your sun exposure. When you're outdoors, use sunscreen.
- If you choose to drink alcohol, do so only in moderation.
- Consult your doctor for regular cancer screenings.
- Reduce exposure to potential cancer-causing substances (carcinogens), such as radon, asbestos, radiation and air pollution.

## No.3—Injuries

The leading cause of fatal accidents among men is motor vehicle crashes, according to the CDC. To reduce your risk of a deadly crash:
- Wear your seat belt.
- Follow the speed limit.
- Don't drive under the influence of alcohol or any other substances.
- Don't drive while sleepy.

Falls and poisoning are other leading causes of fatal accidents. Take common-sense precautions, such as using chemical products only in ventilated areas, using nonslip mats in the bathtub and placing carbon monoxide detectors near the bedrooms in your home.

## No.4—Stroke

You may not be able to control some stroke risk factors, such as family history, age and race. But you can control other contributing factors. For example:
- Don't smoke.
- If you have high cholesterol or high blood pressure, follow your doctor's treatment recommendations.

- Limit the amount of saturated fat and cholesterol (animal protein e.g. meat, cheese, eggs and milk) in your diet. Try to avoid trans-fat entirely.
- Maintain a healthy weight.
- Include physical activity in your daily routine.
- If you have diabetes, keep your blood sugar under control.
- If you choose to drink alcohol, do so only in moderation.

### No.5—COPD

Chronic obstructive pulmonary disease (COPD) is a group of chronic lung conditions, including bronchitis and emphysema. To prevent COPD:
- Don't smoke. Avoid exposure to secondhand smoke.
- Minimize exposure to chemicals and air pollution.

### No.6—Type 2 diabetes

Type 2 diabetes—the most common type of diabetes—affects the way your body uses blood sugar (glucose). Possible complications of type 2 diabetes include heart disease, blindness, nerve damage and kidney damage. To prevent type 2 diabetes:
- Lose excess pounds, if you're overweight.
- Eat a healthy diet rich in fruits, vegetables and low-fat foods.
- Include physical activity in your daily routine.

### No.7—Flu

Influenza is a common viral infection. While a case of the flu isn't usually serious for otherwise healthy adults, complications of the flu can be deadly—especially for those who have weak immune systems or chronic illnesses. To protect yourself from the flu, get an annual flu vaccine.

### No.8—Suicide

Suicide is another leading men's health risk. An important risk factor for suicide among men is depression. If you think you may be depressed, consult your doctor. Treatment is available. If you're contemplating suicide, call for emergency medical help or go the nearest emergency room. You can also call the National Suicide Prevention Lifeline at 800-273-TALK (800-273-8255).

**No.9—Kidney disease**

Kidney failure is often a complication of diabetes or high blood pressure. If you have diabetes or high blood pressure, follow your doctor's treatment suggestions. In addition:
- Eat a healthy diet. Limit the amount of salt you consume.
- Include physical activity in your daily routine
- Lose excess pounds, if you're overweight
- Take medications as prescribed.

**No. 10—Alzheimer's disease**

There's no proven way to prevent Alzheimer's disease, but consider taking these steps:
- Take care of your heart. High blood pressure, heart disease, stroke, diabetes and high cholesterol (animal protein e.g. meat, cheese, eggs and milk) may increase the risk of developing Alzheimer's.
- Avoid head injuries. There appears to be a link between head injury and future risk of Alzheimer's.
- Maintain a healthy weight.
- Include physical activity in your daily routine.
- Avoid tobacco.
- If you choose to drink alcohol, do so only in moderation.
- Stay socially active.
- Maintain mental fitness. Practice mental exercises, and take steps to learn new things.

## Men's Summary

Your bottom line: Take health threats seriously. Health risks can be scary, but there's no reason to panic. Instead, do everything you can to lead a healthy lifestyle—eating a healthy diet, staying physically active, quitting smoking, getting regular checkups and taking precautions in your daily activities. Adopting these preventive measures will increase your odds of living a long, healthy life.

## Women's top 10 Disease Preventions

By Mayo Clinic staff [205]

Many of the leading threats to women's health can be prevented—if you know how. Consider this top 10 list of women's health threats, compiled from statistics provided by the Centers for Disease Control and Prevention (CDC) and other organizations. Then take steps to promote women's health and reduce your risks today.

### No.1—Heart disease

Heart disease isn't just a man's disease. Heart disease is also a major women's health threat. To prevent heart disease:
- Don't smoke or use other tobacco products. Avoid exposure to secondhand smoke.
- Eat a healthy diet rich in vegetables, fruits, whole grains, fiber and fish. Cut back on foods high in saturated fat (animal protein e.g. meat, cheese, eggs and milk) and sodium.
- If you have high cholesterol or high blood pressure, follow your doctor's treatment recommendations.
- Include physical activity in your daily routine.
- Maintain a healthy weight.
- If you choose to drink alcohol, do so only in moderation. Too much alcohol can raise blood pressure.
- If you have diabetes, keep your blood sugar under control.
- Manage stress.

### No.2—Cancer

The most common cause of cancer deaths among women is lung cancer, according to the American Cancer Society. Breast cancer and colorectal cancer also pose major women's health threats. To reduce your risk of cancer:
- Don't smoke or use other tobacco products. Avoid exposure to secondhand smoke.
- Include physical activity in your daily routine.

- Maintain a healthy weight.
- Eat a healthy diet rich in fruits and vegetables, and avoid high-fat (animal protein e.g. meat, cheese, eggs and milk) foods.
- Limit your sun exposure. When you're outdoors, use sunscreen.
- If you choose to drink alcohol, do so only in moderation.
- Consult your doctor for regular cancer screenings.
- Reduce exposure to cancer-causing substances (carcinogens), such as radon, asbestos, radiation and air pollution.
- Breast-feed, if you can.

## No.3—Stroke

You can't control some stroke risk factors, such as age, family history, sex or race. But you can take these steps to reduce your risk of stroke:
- Don't smoke.
- If you have high cholesterol or high blood pressure, follow your doctor's treatment recommendations.
- Limit the amount of saturated fat and cholesterol (animal protein e.g. meat, cheese, eggs and milk) in your diet. Try to avoid trans-fat entirely.
- Maintain a healthy weight.
- Include physical activity in your daily routine.
- If you have diabetes, keep your blood sugar under control.
- If you choose to drink alcohol, do so only in moderation.

## No.4—COPD

Chronic obstructive pulmonary disease (COPD) is a group of chronic lung conditions, including bronchitis and emphysema. To prevent COPD:
- Don't smoke. Avoid exposure to secondhand smoke.
- Minimize exposure to chemicals and air pollution

## No.5—Alzheimer's disease

There's no proven way to prevent Alzheimer's disease, but consider taking these steps:
- Take care of your heart. High blood pressure, heart disease, stroke, diabetes and high cholesterol (animal protein e.g. meat, cheese, eggs and milk) may increase the risk of developing Alzheimer's.

- Avoid head injuries. There appears to be a link between head injury and future risk of Alzheimer's.
- Maintain a healthy weight.
- Include physical activity in your daily routine.
- Avoid tobacco.
- If you choose to drink alcohol, do so only in moderation.
- Stay socially active
- Maintain mental fitness. Practice mental exercises, and take steps to learn new things.

## No.6—Injuries

The leading cause of fatal accidents among women is motor vehicle crashes, according to the CDC. To reduce your risk of a deadly crash:
- Wear your seat belt.
- Follow the speed limit.
- Don't drive under the influence of alcohol or any other substances.
- Don't drive while sleepy.

Falls and poisoning also pose major women's health threats. Take common-sense precautions, such as having your vision checked, using nonslip mats in the tub and placing carbon monoxide detectors near the bedrooms in your home.

## No. 7—Type 2 diabetes

Type 2 diabetes—the most common type of diabetes—affects the way your body uses blood sugar (glucose). Possible complications of type 2 diabetes include heart disease, blindness, nerve damage and kidney damage. To prevent type 2 diabetes:
- Lose excess pounds, if you're overweight.
- Eat a healthy diet rich in fruits, vegetables and low-fat foods.
- Include physical activity in your daily routine.

## No.8—Flu

Influenza is a common viral infection. While a case of the flu isn't usually serious for otherwise healthy adults, complications of the flu can be deadly—especially for those who have weak immune systems or chronic illnesses. To protect yourself from the flu, get an annual flu vaccine.

### No.9—Kidney disease

Kidney failure is often a complication of diabetes or high blood pressure. If you have diabetes or high blood pressure, follow your doctor's treatment suggestions. In addition:
- Eat a healthy diet. Limit the amount of salt you consume.
- Include physical activity in your daily routine.
- Lose excess pounds, if you're overweight.
- Take medications as prescribed.

### No. 10—Blood Poisoning (Septicemia or Sepsis)

Septicemia is a life-threatening infection marked by the presence of bacteria or their toxins in the blood. Septicemia commonly arises from infections in the lung, urinary tract, abdomen or pelvis. Often, it isn't preventable—but you can take steps to avoid infections and to protect yourself from illnesses that weaken your immune system:
- Wash your hands often.
- Keep your vaccines current.
- Seek prompt medical care for any serious infection.
- Change tampons at least every six to eight hours and avoid using superabsorbent tampons.
- Wipe from front to back after urinating and urinate after sex.

## Women's Summary

Your bottom line: Take health threats seriously. It's important to understand common women's health risks, but don't feel intimidated. Instead, do whatever you can to lead a healthy lifestyle—including eating healthy foods, staying physically active, getting regular checkups and paying attention to your environment. Preventive measures can go a long way toward reducing your health risks.

# Appendix F:
# Famous Vegetarians

If you search the web for famous vegetarians you will find many from every country in the world. This is a very short list of people that are vegetarian or Vegan. Some have become so recently.

- Abraham Lincoln
- Albert Einstein
- Aristotle
- Barack Obama
- Ben Franklin
- Bill Clinton
- Brigitte Bardot
- George Bernard Shaw
- HG Wells
- Jesus Christ
- John the Baptist
- Leo Tolstoy
- Leonardo Da Vinci
- Mark Twain
- Martin Luther
- Mary Shelly
- MK Gandhi

- Plato
- Sir Isaac Newton
- Socrates
- St Francis of Assisi
- Steve Jobs
- Virgil
- Voltaire
- William Shakespeare

Some of these people may sway away at times but mainly follow a vegetarian diet.

# Glossary of Terms

**A**

Amino Acids—Basic building blocks of all protein and therefore all life on Earth.

**D**

DNA—Deoxyribonucleic acid DNA is a nucleic acid that contains the genetic instructions that is the basis for all known life on Earth.

**E**

Epigenetic Genome (epigenome)—The study of which genes are expressed or blocked (turned ON/OFF). Epigenetics—is a relatively new area of biology that is changing everything we thought we knew about life. The term epigenetics is "changes to the observable characteristic or trait of an organism. It causes gene expression by mechanisms other than changes in the underlying DNA sequence, hence the name epi—(Greek: over; above) genetics. These changes may remain through cell divisions for the remainder of the cell's life and may also last for multiple generations. However, there is no change in the underlying DNA sequence of the organism; instead, non-genetic factors cause the organism's genes to behave differently".

Epigenome—{see epigenetic genome}

Eukaryote—A cell that contains a nucleus e.g. all cells in human body.

**G**

Gene—A group of DNA codes that act like a memory or program to create a particular protein in our body.

Genome—the genome is the collection of genes that make up an organism's hereditary information. It is usually made of DNA but can be RNA for viruses.

## M

Methyl Groups—cause epigenetic triggers. A "methyl" group is simply one carbon connected to three hydrogen atoms. It may be written as $CH_3$.

Methylation—is not just one specific reaction. There are hundreds of "methylation" reactions in the body. Methylation is simply the adding or removal of the methyl group to a compound or other element.

## N

Nocebo—a negative thought having negative actions like making you ill or dying.

Nutrigenomics—is the study of the epigenetic effects of foods on genes. It looks at each expressed gene and which proteins they create to better understand the effects on our health. Nutrigenomics aims to develop rational means to optimize nutrition, with respect to the subject's genetic constitution of a cell, an organism, or an individual.

## P

Phenotype—any *observable characteristic* or trait of an organism. e.g. eyes, hair, scales, etc.

Placebo—a positive thought having positive actions like healing you.

Protein—Made from a chain of amino acids in a particular order and folded in a specific way. Proteins are used by all life as basic building blocks for cells.

## R

Ribosome—A protein element in the cell that builds a protein designated from the RNA strand. A gene, that is expressed, is copied by dividing it in half (to form an RNA strand) and sent out of the cell's nucleus to the cell itself. This half strand of DNA (RNA) is used to assemble amino acids in a particular order (dictated by the gene DNA sequence). This is a process similar to building a necklace from different colored beads. The RNA calls for a red bead, two yellow beads, and a blue bead and so on until the sequence is complete. The sequence of amino acids forms a particular protein the body needs and uses.

RNA—A half strand copy of a DNA gene used as input to a ribosome to assemble a sequence of amino acids into a protein.

## U

Ubiquitin—A protein that acts like an antibody and aids the immune system.

# Foot Notes

1. Medicinenet.com. "The Definition of Epigenetics"
2. Medicinenet.com. "Definition of Nutrigenomics"
3. Newsweek. March 19, 2007
4. Dr. Lipton, Bruce H. "The Biology of Believe: Unleashing The Power Of Consciousness, Matter And Miracles" Hay House Inc. September, 2008
5. Moll, Rob. "The Art of Dying". IVP Books. 2010
6. Dr. Bruce Lipton. The Magnetic Centre. "The Biology of Belief: An Epigenetic Primer".
7. Fossil Humanoids FAQs. "Fossil Hominids: mitochondrial DNA". http://www.talkorigins.org/faqs/homs/mtDNA.html
8. Dr. Bruce Lipton. YouTube. "Epigenetics: Your unlimited potential for health". http://www.youtube.com/watch?v=a12fzb9ZJ9E
9. Shenk, David. 'The Genius in all of us". Doubleday, 2010. Pg 129
10. Shenk, David. 'The Genius in all of us". Doubleday, 2010. Pg 131
11. Shenk, David. "The Genius in all of us". Doubleday, 2010. Pg 104
12. Shenk, David. "The Genius in all of us". Doubleday, 2010. Pg 19
13. Shenk, David. "The Genius in all of us". Doubleday, 2010. Pg 21
14. Wikipedia. "Human Genome Project"
15. University of Utah, *http://learn.genetics.utah.edu/content/epigenetics/epi_learns/*
16. Watters, Ethan. "DNA is not Destiny". Published online
17. Watters, Ethan. "DNA is not Destiny". Published online
18. Watters, Ethan. "DNA is not Destiny". Published online
19. Neil DeGrasse Tyson. NOVA Science Now. Aug 2007. "Epigenetics—Ask the Expert".
20. Jirtle, Randy Professor. Duke University. Nova Science Now. "Epigenetics". Pbs.org/nova/sciencenow
21. Dr. Bruce Lipton. The Magnetic Centre. "The Biology of Belief: An Epigenetic Primer".

22. PBS NOVA. "NOVA scienceNOW: Epigenetics" *http://www.pbs.org/wgbh/nova/teachers/programs/3411_02_nsn.html*
23. YouTube. "The Epigenome at a Glance". Learn Genetics.
24. Enzyme Stuff. "Methylation". *http://www.enzymestuff.com/methylation.htm*
25. Medical Food News. "What is Omega-3?". *http://www.medicinalfoodnews.com/vol13/omega-3* Oct 2009 # 228
26. Science Daily. "Epigenetics News". http://www.sciencedaily.com/news/plants_animals/epigentics/
27. Wallis, Paul. Digital Journal. "Epigenetics: A possible 'off switch' for diseases". October 6, 2009
28. Bionity.com. "Unlocking the secrets of cellular energy holds promise for obesity, diabetes and cancer".
29. Dr. Bruce Lipton. YouTube. "Epigenetics: How Does It Work".
30. Nicolaus, Martin. New Recovery BLOG. "Goodbye Genetics, Hello Epigenetics". November 7, 2009
31. Bionity.com. "Largest ever epigenetics project launched".
32. Avery M.D. FACP, Robert. "Cancer, Epigenetics, and Nutrigenomics—How Food Affects Your Cancer Genes"
33. Dr. Bannister, Andy. "Key epigenetic processes & links to cancer". Cambridge University
34. Rothstein, Mark A.; Cai, Yu; Marchant, Gary E. "The ghost in our genes: legal and ethical implications of epigenetics". Health Matrix. Jan 1, 2009.
35. MACRAE, Fiona. "Mothers-to-be who gorge on junk food could be putting their grandchildren at risk of breast cancer". Georgetown University
36. Feinberg, Andy, director of Johns Hopkins Epigenetics Center. "Nature meets Nurture"
37. Walsh, ND, Robin. "Nutrigenomics". May 1, 2010.
38. Nutrigenomics: the science of individualized nutrition. http://nutrigenomika.com/
39. Glass, Don. "Nutrigenomics". February 10, 2004. http://indianapublicmedia.org/amomentofscience/nutrigenomics/
40. Environmental Health News. " Chemicals can turn genes on and off; new tests needed, scientists say". *http://www.environmentalhealthnews.org/ehs/news/epigenetics-workshop*
41. YouTube. "The Gene Expression".
42. YouTube. "The Epigenome at a Glance". Learn Genetics.
43. Weil, MD, Andrew. "Eating Well For Optimum Health". Knopf.2000. Pg 60-61
44. Shenk, David. "The Genius in all of us". Doubleday, 2010. Pg 128

45. Robbins, John. "Diet for a New America".
46. Robbins, John. "Diet for a New America". Pg 177
47. Chopra, Deepak. "Journey into Healing". Crown Publishing Group. March 1995
48. Kessler, David A. "The End of Overeating". Rodale Press, Inc. April 2009
49. Khalsa, MD, Dharma Singh. "Food as Medicine". Atria Books. Pg 15
50. Meals Matter. "Making Sense of Portion Sizes". http://www.mealsmatter.org/EatingForHealth/Topics/Healthy-Living-Articles/Portion-Sizes.aspx
51. Kessler, MD, David A. "The End of Overeating". Rodale. 2009
52. Denise Grady. New York Times. "Obesity rates keep rising, troubling health officials". Aug 3, 2010
53. THE BIOLOGY OF BELIEF: An Epigenetic Primer. *http://www.danbartlett.co.uk/lipton_epigenetics.htm*
54. Weil, MD, Andrew. "Eating Well For Optimum Health". Knopf.2000. Pg 48
55. NPR Apr 22,2010—*http://www.npr.org/templates/story/story.php?storyId=126155471&ft=1&f=1003*
56. Medical News Today—*http://www.medicalnewstoday.com/articles/186127.php*
57. Zinczenko, David. "Mens Health". Posted on Mon, May 10, 2010, 11:06 am PDT. *www.health.yahoo.com/nutrition-overview/*
58. Edelbaum, Michelle. Editor for Eatingwell Media Group. "Beware: Misleading ingredient names explained ". September 24, 2010 in Yahoo Food
59. Asheville Citizen-Times—Mar 29 9:26 PM
60. Joanne Slavin of the University of Minnesota. "Nutrition Action Health Newsletter". www.cspinet.org/nah/wwheat.html
61. Rachael Ray. "The Food Network TV or Internet" *http://www.foodnetwork.com/30-minute-meals/index.html*
62. Sage Medical Lab. *http://www.foodallergytest.com/illnesses.html*
63. Bonnie Minsky MA, MPH, LDN, CNS & Steve Minsky Nutritional Concepts, Inc. "Using epigenetics to prevent chronic disease: Part One". 2007.
64. Zap, Claudine. "Have your burgers many ways: burger wars heat up". August 19, 2010. Yahoo News
65. Barnett, Dr. Matt. "The Epigenome Song". 2009 YouTube: *http://www.youtube.com/watch?v=D_YKiXI7l9c*
66. Khalsa, MD, Dharma Singh. "Food as Medicine". Atria Books. Pg 13
67. Khalsa, MD, Dharma Singh. "Food as Medicine". Atria Books.
68. FC&A Medical Publishing. "Eat and Heal". FCA. 2001
69. Church, Dawson. "The Genie in your Genes". Energy Psychology Press & Elite Books. 2007. Pg 101

70. Stoddard, Alexandra. "You are your choices". Collins 2007. Page 168
71. Byrne, Rhonda. "The Secret". Beyond Words Publishing & Atria Books.
72. Byrne, Rhonda. "The Secret". Beyond Words Publishing & Atria Books. Pg 125
73. Byrne, Rhonda. "The Secret". Beyond Words Publishing & Atria Books. Pg 127
74. Byrne, Rhonda. 'The Secret". Beyond Words Publishing & Atria Books. Pg 128
75. Stoddard, Alexandra. "You are your choices". Collins. 2007
76. Stoddard, Alexandra. "You are your choices". Collins. 2007
77. Wikipedia. "Pessimism"
78. Gail's Story. www.rainbowblessings.org.
79. Flanagan, Beverly. "Forgiving the Unforgiveable". Macmillan. 1992
80. Wilson, Edward O. "From Ants to Ethics: A Biologist Dreams Of Unity of Knowledge". NY Times article By NICHOLAS WADE Published: May 12, 1998
81. Wilson, Edward O. "From Ants to Ethics: A Biologist Dreams Of Unity of Knowledge". NY Times article By NICHOLAS WADE Published: May 12, 1998
82. Rubin, Gretchen. "The Happiness Project". Harper. 2009
83. Church, Dawson. "The Genie in your Genes". Energy Psychology Press & Elite Books. 2007
84. Church, Dawson. "The Genie in your Genes". Energy Psychology Press & Elite Books. 2007. Pg 71
85. SRI Narasingha Chaitanya Ashram. "Real Religion is not man-made"
86. Freeman, Morgan (Science Channel). "Through the Wormhole"
87. Stoddard, Alexandra. "You are your choices". Collins 2007. Page 147
88. Pppst.com. http://science.pppst.com/fivesenses.html
89. Koontz, Dean. "Frankenstein Lost Souls". Bantam 2010 Page 312 of hardcover
90. Wikipedia. www.wikipedia.org Definition of Vegetarianism
91. "K", Dr. Betty. "Vegetarian Cuisine". March 1993. Betty K Books & Food
92. GoVeg.com. http://www.goveg.com/healthConcerns.asp
93. Barnouin, Kim. Skinny Bitch—Ultimate Everyday Cookbook. 2010. Running Press. Page 259
94. Dr Neal Barnard, youtube video at :http://www.youtube.com/watch?v=5VWi6dXCT7I&feature=related", "Chocolate, Cheese, Meat, and Sugar -- Physically Addictive".
95. T. Colin Campbell, PHD. "The China Study" 2005

96  Nicholas Wade. New York Times. Aug 2, 2010. "Breast milk sugars give infants a protective coat"
97  Rory Freedman and Kim Barnouin, "Skinny Bitch", 2005, Running Press
98  The Atma Jyoti Blog—*http://blog.atmajyoti.org/2008/04/humans-are-we-carnivores-or-vegetarians-by-nature/*
99  The Atma Jyoti Blog—*http://blog.atmajyoti.org/2008/04/humans-are-we-carnivores-or-vegetarians-by-nature/*
100 Freston, Kathy. "Shattering the Meat Myth: Humans are natural vegetarians". June 11, 2009. Author, Health & Wellness Expert
101 Mills, MD, Milton R. "The Comparative Anatomy of Eating". Nov 21, 2009.
102 Homo sapiens. *http://www.esp.org/humor/zoo-lbl.pdf*
103 Furuya, Yukio. "The Relation of Effects of Dietary Changes to Physical and Mental Disorders and Crime Occurrence among Youths". *http://www.jicef.or.jp/wahec/ful418.htm*
104 The Chicago Western Electric Study. "Relation of vegetables, fruit, and meat intake to 7-year Blood Pressure Study in middle aged men. *http://aje.oxfordjournals.org/cgi/reprint/159/6/572.pdf*
105 "Managing Biodiversity in Agricultural Ecosystems". Symposium in Montreal, Canada. Nov 8-10, 2001. *http://www.unu.edu/env/plec/cbd/Montreal/papers/Johns.pdf*
106 HUNG, D.D.S., HSIN-CHIA. "Tooth loss and dietary intake". JADA Continuing Education. *http://jada.ada.org/cgi/content/full/134/9/1185*
107 Vegan Diet Helps Fight Prostate Cancer, Study Says. *http://www.vegtaste.com/pages/posting.php?articleId=200*
108 Posted at 12:21 AM/ET, March 31, 2010 in USA TODAY editorial I Permalink
109 The Center for Disease Control and Prevention
110 T. Colin Campbell, PHD. "The China Study" 2005
111 Health.com. "10 Healthiest Ethnic Cuisines".
112 Yahoo Diet. "The Age Erasing Diet". Friday April 23, 2010
113 Yahoo Diet. "The Age Erasing Diet". Friday April 23, 2010
114 "K", Dr. Betty. "Vegetarian Cuisine". March 1993. Betty K Books & Food
115 Fraser, Linda. "Vegetarian Cooking". 1998, 2003. Barnes & Noble, Inc and Anness Publishing Limited
116 Elliot, Rose. "The Complete Vegetarian Cuisine". 1988, 1996. Pantheon Books New York
117 All 4 natural Health. *http://www.all4naturalhealth.com/benefits-of-vegetarian-diet.html*

118 Free Dallas—Fort Worth Vegetarian Education Network Newsletter. *http://www.dfwnetmall.com/veg/yourfoodandyourhealth.htm*

119 Hirayama, T., "Epidemiology of Breast Cancer with Special Reference to the Role of Diet, *Preventative Medicine 7* (1978): 173-95

120 William Castelli, MD, Director, Framingham Health Study: National Heart, lung, and Blood Institute

121 *Journal of the National Cancer Institute* 92 (2000}:61-8

122 "Dairy Products Linked to Prostate Cancer," *Associated Press,* April 5, 2000

123 Singh, P.N., et aI., "Dietary Risk Factors for Colon Cancer in a Low-Risk Population," American Journal of Epidemiology 148 (1998):761-64

124 1997 the World Cancer Research Fund analyzed more than 4,500 cancer research studies. In their major international report, the analysis concluded the above

125 Position of American Dietetic Association on Vegetarian Diets," *Journal of the American Dietetic Association 97* (1997): 1317-21.

126 Ornish 0, Brown SE, Scherwitz LW, et al. Can lifestyle changes reverse coronary heart disease? *Lancet* 1990;336:129-33.

127 Ophir O., et al., "Low Blood Pressure in Vegetarians…," *American Journal of Clinical Nutrition* 37 (1983):755-62.

128 Ophir O., et al., "Low Blood Pressure in Vegetarians." *American Journal of Clinical Nutrition* 37 (1983):755-62

129 Kokkad, A., et al., "the Spread of the Obesity Epidemic in the United States," *Journal of the American Medical Association* 181 (1999); 1519-22.
Key, T., et al., "Prevalence of Obesity Is Low in People Who Do Not Eat Meat," *British Medical* Journal 313 (1996):816-7

130 Troiano, R., et al., Overweight Children and Adolescents." *Pediatrics 101* (1998) 497-504 and "Overweight Prevalence and Trends for Children and Adolescents," *Archives of Pediatric and Adolescent Medicine* 149 (1995); 1083-91.

131 Halweil, Brian, "United States Leads World Meat Stampede," Worldwide Issues Paper, July 2, 1998.

132 Breslau NA, Brinkley L, Hill KD, Pak CYC. Relationship of animal protein-rich diet to kidney stone formation andcalcium metabolism. *Journal Clinical Endocrinol 1988;66:140-6.*

133 Remer T, Manz F. Estimation of the renal net acid excretion by adults consuming diets containing variable amounts of protein. *American Journal Clinical Nutrition 1994;59:1356-61.*

134 Science Daily. "Lifelong vegetarian diet reduces the risk of colorectal cancer". May 23, 2007

135 Science Daily. "Original human 'Stone Age' diet is good for people with diabetes study finds". June 28, 2007
136 Vegetarianism. BookRags. http://www.bookrags.com/research/vegetarianism-este-0001_0004_0/
137 Julie Steenhuysen. Reuters, Chicago. Monday May 17 at 3:23 PM PDT
138 Mendosa, David, *http://www.mendosa.com/gilists.htm*
139 Weil, MD, Andrew. "Eating Well For Optimum Health". Knopf. 2000. Pg 36
140 Web MD. "The Basics of a Healthy Diabetes Diet" *http://diabetes.webmd.com/diabetes-diet-healthy-diet-basics*
141 Wikipedia. "Diabetes mellitus". *http://en.wikipedia.org/wiki/Diabetes_mellitus*
142 Traverso, Matt. "How to Reverse Diabetes Now!". *http://www.reverse-diabetes-today.com/?hop=china232*
143 Science Daily. "Glycemic Index". http://www.sciencedaily.com/articles/g/glycemic_index.htm
144 Kelly, John, MD. The American College of Lifestyle Medicine. www.aclm.net
145 Studio Press, *http://invisibleillnessweek.com/*.
146 Griffin, G. Edward. *"WORLD WITHOUT CANCER". FTR (Foundation For Truth and Reality-http://www.ftrbooks.net/health/cancer/world_wo_cancer.htm*
147 MARCHIONE, MARILYNN (*Associated* Press). "Americans get most radiation from medical scans". *http://news.yahoo.com/s/ap/20100614/ap_on_he_me/us_med_overtreated_radiation*
148 Editors of FC&A Medical Publishing. *"Eat and Heal".* FC&A Medical Publishing, 2001. Pg 5
149 Gene Zimmer—FTR (Foundation for Truth in Reality)—*http://www.ftrbooks.net/psych/drug_companies.htm*
150 Angell, Marcia. The New York Review of Books. "Drug Companies & Doctors: A Story of Corruption". *http://www.nybooks.com/articles/archives/2009/jan/15/drug-companies-doctorsa-story-of-corruption/*
151 Adams, Chris. "Student Doctors Start to Rebel Against Drug Makers' Influence". Wall Street Journal June 24, 2002
152 Nicole M. Avena, Pedro Rada, and Bartley G. Hoebel. "Evidence for sugar addiction: Behavioral and neurochemical effects of intermittent, excessive sugar intake". National Institute of Health
153 Perkins, M Ed., Cynthia. "The hidden dangers of sugar addiction". No-Hype Holistic Health Solutions.
154 Serge Ahmed, Ph.D. "Intense Sweetness Surpasses Cocaine Reward" *www.plosone.org/article/info:doi/10.1371/journal.pone.0000698*

155. Bart Hoebel, Ph.D. "Sugar can be addictive, Princeton scientist says." *http://www.princeton.edu/main/news/archive/S22/88/56G31/index.xml?section=topstories*
156. familydoctor.org. "Added Sugar: What You Need To Know". *http://familydoctor.org/online/famdocen/home/healthy/food/general-nutrition/1005.html*
157. Healthy Eating Club. "Sugar". *http://www.healthyeatingclub.com/info/articles/diets-foods/sugar.htm*
158. Baldauf, Sarah. "Foods Surprisingly High in Added Sugar". Yahoo Health. *http://health.yahoo.com/featured/35/foods-surprisingly-high-in-added-sugar/*
159. Jefferson, Jamie, "Breaking the Sugar Addiction". Self Growth.com. *http://www.selfgrowth.com/articles/Jefferson14.html*
160. Scott-Thomas, Caroline. "Animal study suggests existence of sugar addiction says scientist". foodnavigator-usa.com
161. Daniells, Stephen. "Food Addiction: Fat may rewire brain like hard drugs". foodnavigator-usa.com
162. Prusiner, M.D., Stanley B. "The Medical News". *http://www.news-medical.net/news/2004/08/02/3710.aspx*
163. Stephanie Relfe B.Sc., Health Wealth & Happiness, *http://www.relfe.com/microwave.html*
164. Mike Adams, Organic Consumers Association, *http://www.organicconsumers.org/articles/article_6463.cfm*
165. Phillips RL. Role of lifestyle and dietary habits in risk of cancer among Seventh-day Adventists. Cancer Res 1975;35(Suppl):3513-22.
166. T. Colin Campbell, PHD. "The China Study" 2005
167. Rory Hafford. Irish Medical Times. "Diagnosis and management of cow's milk protein allergy". *http://www.imt.ie/clinical/paediatrics/diagnosis-and-management-of-co.html*
168. PubMed.gov. "Is type 1 diabetes a disease of the gut immune system triggered by cow's milk insulin?" *http://www.ncbi.nlm.nih.gov/pubmed/16137120?dopt=Abstract*
169. T. Colin Campbell, PHD. "The China Study" 2005
170. Answers.com. "Protein"
171. Disabled World. http://www.disabled-world.com/disability/statistics/aging-diseases.php From Center for Disease Control (CDC) report 2007 State of Aging Report
172. Lauren Cox. Live Science. "Cigarette smoke jolts hundreds of genes, researchers say". Thu Jul 15, 9:55 am ET. www.livescience.com

173 Colleen Trombley-VanHoogstraat. Self Help Authority. "Is Chronic Illness at the Heart of Our Economic Crisis?" *http://www.neuro-vision.us/ad/Article/Is-Chronic-Illness-at-the-Heart-of-Our-Economic-Crisis-/15270*
174 AMFood DNA (Dynamic Nutrition Advantage) from Amega Global
175 Gommer L.Ac., Barry L. "Living Well Magazine". www.livingwellmagazine.com May 2010
176 St. Joseph's Hospital—https://www.stjosephsatlanta.org/HealthLibrary/content.aspx?pageid=P00237
177 St. Joseph's Hospital—http://www.stjosephsatlanta.org/greystone/centers/cancer/overview.html
178 St. Joseph's Hospital—*http://www.stjosephsatlanta.org/HealthLibrary/content.aspx?pageid=P00335*
179 St. Joseph's Hospital—*https://www.stjosephsatlanta.org/HealthLibrary/content.aspx?pageid=P00209*
180 MayoClinic.com—*http://www.mayociinic.com/health/cancer-prevention/ca00024*
181 Membrane.com. "Methyl". *http://ygraine.membrane.com/enterhtml/free/FoodForThought/lyrics/herb/Z08_Methyl.html*
182 MayoClinic.com—*http://www.mayoclinic.com/health/diabetes-prevention/DA00127*
183 MayoClinic.com. "Fever" *http://www.mayoclinic.com/health/fever/DS00077*
184 eHow.com. "How to get rid of Body Aches". *http://www.ehow.com/how_2312188_get-rid-body-aches.html*
185 Maté, Babor. "When the Body Says No". Wiley, John & Sons. April 2003
186 Ornish, MD, Dean. "The Spectrum". Ballantine Books. 2008 Pg. 191-192
187 Ornish, MD, Dean. "The Spectrum". Ballantine Books. 2008 Pg. 185-186
188 Ornish, MD, Dean. "The Spectrum". Ballantine Books. 2008 Pg. 157
189 Ornish, MD, Dean. "The Spectrum". Ballantine Books. 2008 Pg. 162
190 Ornish, MD, Dean. "The Spectrum". Ballantine Books. 2008 Pg. 175
191 Center for Young Woman's Health. "Stress and How to Lower it". *http://www.youngwomenshealth.org/stress.html*
192 Science Daily. "Researchers create animal model of chronic stress". Sept 4, 2008
193 Byrne, Rhonda. "The Secret". Beyond Words Publishing & Atria Books. Pg 12
194 Dobson, Roger. "Death Cab Be Cured". MJF Books, Pg 31
195 Khalsa, MD, Dharma Singh. "Food as Medicine". Atria Books. Pg 8
196 Ornish, MD, Dean. "The Spectrum". Ballantine Books. 2008. Pg 201

197. Stoddard, Alexandra. "You are Your Choices". Collins. 2007
198. Stoddard, Alexandra. "You are your choices". Collins 2007. Page 115
199. Avery M.D. FACP, Robert. "Cancer, Epigenetics, and Nutrigenomics—How Food Affects Your Cancer Genes"
200. YouTube. "The Epigenome at a Glance". Learn Genetics.
201. Halweil, Brian, "United States Leads World Meat Stampede," Worldwide Issues Paper, July 2, 1998.
202. Byrne, Rhonda. "The Secret". Beyond Words Publishing & Atria Books. Pg 12
203. Ornish, MD, Dean. "The Spectrum". Ballantine Books. 2008. Pg 201
204. MayoClinic.com-http://www.mayoclinic.com/health/mens-health/MC00013
205. MayoClinic.com-http://www.mayoclinic.com/health/womens-health/W000014

# Index

addiction, 10, 13, 74, 105, 153, 154, 155, 157, 159, 160, 198, 211, 231, 232
Advertisements, 9, 12, 65, 74, 138, 139, 150, 210
Amino Acids, 8, 14, 34, 41, 46, 55, 63, 64, 73, 74, 113, 114, 115, 121, 161, 162, 163, 164, 168, 179, 203, 223, 224
Arthritis, 59, 78, 113, 119, 167, 169
Belief, 27, 50, 55, 56, 62, 93, 94, 99, 100, 104, 133, 143, 150, 164, 171, 179, 187, 197, 200, 225, 227
Believe, 7, 12, 44, 55, 85, 93, 94, 95, 99, 101, 105, 124, 133, 137, 197, 209, 225
BLOG, 12, 13, 21, 25, 200, 212
Blood Sugar, 8, 79, 129, 130, 131, 132, 158, 178, 192, 213, 215, 217, 218, 219
Body Mass, 73, 94, 130, 193
Breast Cancer, 53, 59, 106, 120, 164, 177, 199, 217, 226, 230
Cancer Research, 6, 54, 59, 171, 230
cholesterol, 24, 65, 69, 70, 109, 119, 148, 154, 180, 213, 214, 215, 216, 217, 218
Chromosomes, 5, 14, 28, 39, 40, 54
Chronic Disease, 79, 111, 143, 144, 167, 198, 227
Chronic Illness, 7, 10, 42, 49, 71, 78, 82, 121, 122, 134, 136, 142, 143, 147, 169, 215, 220, 233
Colon Cancer, 53, 120, 230
Cookbooks, 12, 14, 123, 124, 194, 198
Cooking, 7, 67, 70, 72, 76, 77, 112, 117, 124, 194, 198, 229
Dairy, 58, 67, 72, 102, 103, 120, 122, 128, 199, 208, 230
Dementia, 79, 119, 136, 169
Depression, 79, 143, 154, 155, 160, 168, 169, 186, 192, 215

# Index

Diabetes, 8, 11, 13, 49, 54, 61, 73, 75, 76, 78, 79, 103, 105, 106, 109, 113, 119, 122, 129, 130, 131, 132, 134, 136, 148, 154, 155, 159, 160, 163, 164, 167, 169, 173, 174, 175, 178, 180, 192, 200, 201, 209, 213, 215, 216, 217, 218, 219, 220, 226, 231, 232, 233

DNA, 5, 12, 14, 23, 24, 27, 28, 29, 30, 31, 33, 34, 35, 36, 37, 38, 39, 40, 41, 42, 43, 44, 45, 46, 47, 48, 49, 50, 51, 52, 54, 55, 56, 64, 65, 67, 82, 90, 91, 94, 95, 97, 106, 110, 114, 133, 134, 161, 162, 164, 171, 172, 187, 189, 190, 197, 203, 204, 205, 206, 207, 208, 212, 223, 224, 225, 232, 233

Double Helix, 37

Drug Companies, 9, 136, 139, 148, 149, 150, 151, 152, 180, 189, 201, 231

Enzyme, 52, 53, 79, 108, 165, 226

Epigenome, 6, 31, 46, 49, 50, 52, 54, 55, 61, 82, 143, 198, 223, 226, 227, 234

Eukaryote, 50, 223

Evolution, 10, 14, 35, 42, 109, 114, 122, 133, 162, 163, 164, 165, 205, 206, 211

Evolutionary, 10, 114, 162, 163, 164, 165, 211

Exercise, 36, 65, 73, 79, 92, 94, 120, 134, 136, 142, 145, 147, 148, 150, 151, 152, 156, 174, 180, 183, 189, 190, 195, 209, 216, 219

Famous Vegetarians, 13, 221

Fat, 6, 36, 38, 39, 41, 42, 45, 48, 49, 53, 58, 59, 63, 65, 68, 69, 70, 72, 73, 74, 75, 78, 83, 104, 105, 109, 110, 111, 112, 113, 116, 118, 120, 122, 124, 128, 130, 131, 135, 146, 147, 148, 151, 154, 157, 159, 168, 169, 175, 176, 177, 180, 190, 191, 192, 193, 198, 201, 207, 208, 213, 214, 215, 217, 218, 219, 232

Fish, 10, 17, 24, 58, 66, 71, 72, 73, 91, 102, 103, 109, 111, 113, 114, 122, 145, 163, 164, 176, 191, 194, 208, 213, 217

Forgive, 87, 88, 198, 228

Forgiving, 7, 87, 88, 195, 198, 228

Fruit, 7, 10, 33, 39, 47, 49, 53, 58, 65, 66, 67, 68, 71, 72, 73, 74, 75, 76, 94, 95, 102, 103, 105, 106, 107, 108, 109, 110, 111, 112, 114, 115, 117, 120, 121, 122, 123, 125, 126, 127, 128, 134, 135, 139, 144, 145, 150, 151, 153, 154, 155, 157, 158, 163, 165, 168, 170, 171, 176, 177, 178, 189, 191, 192, 193, 194, 199, 207, 208, 213, 214, 215, 217, 218, 219, 229

Fruit Juice, 7, 75, 155, 158, 176, 189
Gene Expression, 28, 54, 59, 168, 198, 223, 226
Genome, 5, 23, 36, 46, 49, 52, 54, 59, 79, 162, 223, 225
Genome Project, 5, 36, 162, 225
Glycemic Index, 8, 12, 14, 80, 126, 127, 129, 130, 131, 153, 160, 192, 209, 231
God, 7, 14, 32, 55, 67, 75, 80, 81, 91, 93, 94, 95, 96, 98, 99, 100, 105, 106, 134, 135, 136, 149, 151, 188, 196
Grains, 7, 58, 76, 102, 108, 112, 131, 176, 177, 178, 208, 213, 217
Happiness, 9, 31, 44, 83, 86, 88, 89, 91, 92, 100, 110, 145, 150, 188, 189, 228, 232
Heart Disease, 8, 11, 13, 61, 73, 74, 76, 78, 103, 104, 109, 111, 113, 119, 120, 122, 132, 134, 154, 163, 164, 167, 169, 172, 173, 175, 178, 209, 212, 213, 215, 216, 217, 218, 219, 230
High Blood Pressure, 8, 24, 65, 119, 120, 122, 148, 154, 155, 172, 180, 213, 214, 216, 217, 218, 220
High cholesterol, 24, 119, 154, 213, 214, 216, 217, 218
Hypertension, 79, 119
Identical Twins, 24, 45, 47, 48, 49, 50
Inheritance, 6, 44, 51, 204
Kinase Hubs, 79
Leukemia, 53
Mammals, 29, 34, 35, 95, 106, 205
Marketing, 9, 72, 73, 100, 110, 133, 136, 138, 202, 210
Meat, 10, 12, 14, 17, 58, 66, 72, 73, 74, 77, 83, 100, 102, 103, 104, 105, 106, 107, 108, 109, 110, 111, 113, 114, 115, 118, 119, 120, 121, 122, 123, 126, 145, 162, 163, 164, 165, 176, 180, 190, 191, 194, 195, 199, 208, 209, 211, 213, 214, 215, 216, 217, 218, 228, 229, 230, 234
Medical Costs, 8, 121, 209
Medical Schools, 9, 149, 150
Meditation, 54, 92, 124, 138, 145, 147, 180, 183, 184, 192, 209
Methyl Groups, 6, 46, 51, 52, 53, 63, 177, 224
Methylation, 49, 51, 52, 53, 54, 58, 82, 208, 224, 226
Milk, 8, 10, 66, 74, 104, 105, 106, 111, 113, 128, 156, 159, 164, 176, 189, 191, 193, 198, 200, 213, 214, 215, 216, 217, 218, 229, 232
Misinformation, 9, 133

# INDEX

Mood, 31, 43, 113, 154, 155, 158, 179, 183
Natural Selection, 5, 35
Nutrition, 10, 31, 45, 60, 61, 64, 67, 69, 79, 121, 124, 131, 147, 151, 155, 162, 168, 169, 173, 198, 199, 200, 224, 226, 227, 230, 232, 233
Nuts, 33, 39, 58, 72, 74, 102, 112, 113, 117, 122, 123, 157, 158, 163, 176, 178, 192, 193, 194, 207, 208
Obese, 71, 73, 103, 121, 136, 139, 140, 142, 147, 156, 201
Obesity, 8, 49, 54, 61, 71, 103, 105, 106, 109, 119, 120, 134, 135, 136, 147, 154, 155, 160, 163, 169, 177, 199, 226, 227, 230
Osteoporosis, 8, 103, 119, 121
Phenotype, 30, 224
Prayer, 94, 183, 184
Prions, 10, 161, 162
Prostate Cancer, 53, 119, 120, 199, 214, 229, 230
Protein, 7, 8, 10, 13, 14, 28, 30, 31, 34, 36, 37, 41, 42, 45, 46, 50, 52, 54, 55, 64, 66, 67, 68, 70, 73, 74, 75, 90, 100, 106, 111, 113, 114, 115, 116, 117, 121, 123, 131, 134, 151, 158, 159, 161, 162, 163, 164, 165, 168, 176, 189, 190, 191, 200, 202, 203, 205, 211, 213, 214, 215, 216, 217, 218, 223, 224, 230, 232
Qigong, 183
Religion, 7, 14, 56, 94, 95, 99, 100, 228
Ribosome, 31, 46, 73, 203, 224
RNA, 46, 52, 73, 161, 203, 205, 223, 224
Senses, 7, 14, 45, 96, 97, 98, 205, 228
Skin Cancer, 53
Smoke, 6, 12, 50, 63, 77, 78, 100, 160, 168, 179, 189, 200, 208, 213, 214, 215, 217, 218, 232, 234
Starvation Mode, 38, 39, 65, 71, 135, 190, 191, 207
Stress, 11, 13, 25, 30, 33, 37, 42, 46, 47, 50, 54, 56, 62, 64, 71, 78, 79, 80, 85, 86, 90, 94, 95, 99, 100, 120, 125, 129, 130, 134, 136, 137, 143, 145, 147, 148, 149, 170, 171, 178, 179, 180, 181, 182, 183, 184, 185, 186, 187, 189, 190, 208, 209, 211, 212, 213, 217, 233
Stroke, 13, 73, 103, 113, 119, 172, 173, 175, 214, 216, 218
Tai Chi, 183
Tobacco, 10, 140, 168, 177, 213, 214, 216, 217, 219

Vegan, 102, 103, 104, 110, 118, 121, 123, 124, 163, 191, 194, 199, 221, 229
Vegetarian, 8, 12, 13, 24, 74, 102, 103, 110, 111, 114, 117, 118, 119, 120, 121, 122, 123, 148, 161, 163, 165, 180, 189, 190, 199, 202, 209, 211, 221, 222, 228, 229, 230, 231
Vegetarianism, 8, 12, 102, 119, 121, 122, 161, 202, 209, 228, 231
Vitamins, 9, 66, 76, 111, 147, 162
Weight, 8, 11, 39, 61, 67, 69, 71, 73, 75, 94, 109, 115, 120, 121, 130, 132, 134, 135, 146, 150, 173, 174, 177, 178, 180, 190, 191, 192, 193, 194, 195, 207, 213, 214, 215, 216, 217, 218, 219, 220, 230
Wine, 71, 80, 113
Womb, 6, 59
Yoga, 54, 151, 183, 184, 192

Edwards Brothers, Inc.
Thorofare, NJ USA
September 16, 2011